Nature of Reality

Nature of Reality

Nature of Reality

A personal reflection on several epistemological concepts

Sattvik Basarkod

Nature of Reality

Copyright © 2022
All rights reserved by Sattvik Basarkod

Dedication

To my dear mother, father, my sister
and my high school TOK teacher, Mr. Shankles,
and to all the millions of high school students
with a passion for philosophy and epistemology.
Without their support, this task wouldn't have been possible.

Nature of Reality

Table of Contents

01.	Paradox of Language	9
02.	Knowledge	14
03.	Sense Perception	20
04.	Concept of Education	28
05.	Concept of Science	40
06.	Concept of Arts	54
07.	Politics and Economics	62
08.	Concept of Emotion	71
09.	Concept of Compassion	77
10.	Concept of Memory	86
11.	Concept of God and Free Will	101
12.	Enlightenment	119
13.	Preface	126
14.	About The Author	128
15.	References	129

Nature of Reality

1. Paradox of Language

Language and Thought

When seeking to understand the nature of knowledge, it is important to analyze how knowledge is gained. The problem of knowledge is the problem of language. So long as there is a structural way of thinking via linguistic methods, knowledge is predisposed to be structured in a similar way. The problem with language is that we can't use language to break down language. We can't escape to a realm of thought that transcends language. For example, read this word: CAT. But now read this word: KAT. Beyond the phonetics and the way the letters are arranged, the language implicates a different thought when reading CAT vs KAT. We think of language and thought as the same. What comes first though? Is it that thoughts in our head precede the communication of ideas via language? Or does language dictate the process of thought, and those thoughts then create a certain image in the head. Language cannot be examined through its medium. That is like breaking down a house and using the wreckage to build a completely new house. The new house just won't be new, nor different.

Oftentimes, language is the easiest medium of understanding. Certainly, defining and thinking about knowledge requires language. Yet if language is meaningless, in the sense that it is only a creation and not an identification method, then the whole chain of knowledge and seeking it becomes a creation as well. All that you have read so far seems like a process of mind reading. How is it that thoughts come down to paper and convey their meaning? Or perhaps they create an entirely new meaning, leaving us in hindsight bias, blind to the reality of the original thought. Writing is very similar to art. When looking at an art piece, there is an interaction between the knowledge of the art, and the knowledge perceived in the head. The art piece itself isn't knowledge, it is simply a method of attaining knowledge. Poems and songs in themselves do not contain some essence, it is the way they are presented through literary devices that gives them their meaning Thus, language and thought act in similar ways. Language is the medium that people use to think. The mediator and the act of mediation are separate for the mediator acts only to elicit a quick response that in totality is called the act of mediation.

Communication

Since thought and language are separate, it can be concluded that whatever I am typing on my computer is different from the essence of my thoughts. This is the problem of communication. There are two types of communication errors: the error of communicating within yourself, and the error of communicating with others. The act of communication refers to an interaction between the communicator and the receiver of the communication. Yet words and languages are stagnant. They don't talk back to you. The same can be said for understanding. Understanding means that the input is in a structure that is compatible with the output's receiver's understanding. One cannot talk to a bear by using sign language or any other verbal language. Communication never occurs if both parties are not using the same means of understanding one another. So why do so many people think writing or reading is understanding? Isn't it that the words themselves aren't conveying any message, rather it is you making a conception of their meaning. In terms of compatibility, the processes within you and the processes of the language are never compatible. So then how do we understand? People prefer external sources that may not be compatible with their understanding rather than internal sources of intuition that are more compatible with the individual's thoughts and actions. But here's the problem: Intuition and internal sources are created based on external sources. This cycle of reinforcement of thought and knowledge regarding language and communication keeps us distant from the truth. The truth about what... can it even be communicated? Can it ever be distinct? Or is all and any knowledge paradoxical in the linguistic and communicative methods?

Paradox of Language

Answering that by using words would be ironic. So that will be for you to decide.

Let's take a step back from the nature of communication and look at the history of language. As humans, we are told and believe that we are social creatures. We have 'evolved' in a way that suggests people to be social. In this manner, language is created. Understanding is the goal, and language is the method. The first way of communication was by drawing on caves. This shows that language wasn't intuitively beneficial but was used to produce efficiency. Regardless of that fact, the drawings communicated images in the caveman's heads. This produced a stimulus for the cavemen that then caused them to act in a certain way. Was this voluntary or autonomic? Well, Darwin's theory would argue that this was autonomic since people evolved in a way to benefit their survival chances. And the communication was the best way to do that. But it could've been voluntary as a method of finding more about the world and people around. A consensive conclusion cannot be drawn because history is simply a thought in our heads now.

About 100,000 years after the drawings, language and vocal communication was invented. Humankind was able to output a vocal voice with an intent that was compatible with the voice's message in the receiver's head. We conditioned ourselves to believe that what we conceive the message to be is the actual intent of the messenger. That is one of the reasons why babies can't differentiate between lies and truth. Babies haven't conditioned themselves to understand that the voice in someone else's head could be different than the voice in their own head. Suspicion could be argued to be the worst evil. Because that is what gave rise to wars and other disasters. It also helped Freud invent his psychosexual theory of ego and the superego. Anyways, going back to the concept of communication, we can, at last, confirm that humans created a language system. Language being a social construct confirms the fact that it itself has many limitations. And as humans, who are communicative and social creatures, we are bound to be entrapped in those limitations. I cannot communicate the limitations of communication via communicating. That's how concrete the paradox is! And that is the message! It isn't the finger that I want you to look at, but rather the thing that is being pointed by the finger. If at any time, the language seems to be difficult to comprehend and perhaps meaningless, remember that language is only the means. It is only the finger. The thought that arises from this communication is the thing you should be looking for.

Nature of Words and Meaning

Let's go a little deeper into the meaning of meaning. The production of language is very complicated and the specific systems of the body and the brain that correspond to language is a topic that will be examined later in this book. Yet, we haven't really asked ourselves what language truly is. The basis of language relies on the fact of distinction. Each word doesn't contain its own meaning so long it has a contrasting partner. A cat is only a cat so long I can differentiate it from other animals not named cat. Similarly, any object in the world has its own identity based on the identity around that individual object. Thus language too contains essence so long there are words that distinguish matter and objects. This phenomenon is known as individualization. Each object needs its opposite object to have a meaning. A cat won't be a cat unless there are other animals that do not resemble a cat. In these ways, we are able to differentiate and come up with meaning. Since we have concluded that language is a social construct, we can also conclude that the act of identification and differentiation is also just a construct to aid our understanding. And if differentiation is a construct then that means that the reality of knowledge lies in deconstructing that construct. Which then creates a possibility of there being absolutely nothing. Destructing the differentiation practically means not understanding, or perhaps not creating boundaries around the form or meaning of the object. Ultimately deconstructing the meaning and language leads to the construction of nothingness. In other words, language helps us not fall into an endless well. It allows us to produce certain boundaries and create meaning. Without language, the connotation of objects wouldn't exist, nor would the world, and people would live meaninglessly. Yet would exist independently, but just not in the shared reality. The bottom line is: that we need language to construct reality, So then if it is constructed, will it ever be real? Again going back to the language problem, we can't answer that. If anything, the answer would be a construction of language and communication as well, wouldn't it?

This Book

Now that I have presented the flaws and limitations of language, as well as its helpful use, I can explain how I will solve this language problem in this book. Clearly, the only way of communicating is by writing in today's day and age. Mind reading and clairvoyance practices are not in the realm of true knowledge. The way I would approach this book is by reading and thinking. Clearly, not all words in this book will correspond to the essence of that topic in my head. Thus use the language and the content in the book as a means of understanding and learning. Don't get caught up in the language of it, but examine the thoughts that are provoked after reading it. Again, like everyone else, I can only put forth what language permits me, but I am hoping that language guides the reader to think out of the box and gain an understanding of reality and truth.

2. Knowledge

Perception vs Reality

The problem of knowledge derives from the concept of perception. No matter how real any knowledge might be, if it isn't perceived as real, it is not real in our realm. Similarly, no matter how unreal any knowledge might be, if it is perceived as real in our realm, there is no way to label it as unreal knowledge. This problem concludes that all knowledge is perceived to be real or unreal and perception does not necessarily define reality, although it does a good job of hiding it.

As humans, we are predisposed to be in the realm of perception. This doesn't mean that our knowledge is uncertain, although it could be, it means that in order to distinguish perception and reality one has to look at them from a third person's perspective. One cannot try to distinguish perception and reality in the realm of perception. The only way to escape this inevitable realm of perception is through meditation and mindfulness.

Meditation is not part of a religion, or any philosophy nor does it have its own category. Meditation is a way of life. In order to escape the problem of perception, one has to unlearn before learning. All our lives, we have been using perception to define reality. This happens with science, history, reasoning, and all other areas of knowledge. By meditating, we are reprogramming our minds to think in a nonlinear and non-perceptive way. It is here that you will be able to distinguish the deception of perception This approach relies on the metaphysical theory of reality as it postulates knowledge in the mind is different from knowledge that is perceived. Those who think that perception is a reality often deceive themselves. Science and perception can only go so far in terms of defining or understanding certain knowledge. Quantum physics itself is in the process of becoming pseudoscientific. Yet these matters will be discussed later on in the book. Focusing on knowledge itself, it can be concluded that in order to distinguish reality and perception, one has to meditate and escape the realm of perception.

Affirmation of Knowledge

Perception is often timed and shaped by a collective society. From the day we are born, our knowledge depends on what society chooses to offer and not offer. Knowledge, at least in the realm of perception, is seen as more real if it is confirmed by others around. The entities around us produce a strong impact on our say on what is real and nonreal. Even a lion thinks of itself as a sheep, if it is born among sheep. The lion eats like sheep, drinks like sheep, and sleeps like sheep. It is only when it sees its reflection in the mirror that the lion realizes its whole life has been a lie.

Science evolves over time with technology, and so does our perception of reality. That then brings up the question of whether knowledge is dependent on the time period? Is knowledge ever so evolving or is it something stagnant that has been there for as long as the universe? Was there anything before the universe? Was knowledge of the universe there before the universe? Does knowledge need a space of realm, or can it exist without space and entity? If knowledge does evolve, hypothetically, then is it either evolving by itself with reference to time and space, or it is evolving by the means of the seeker who evolves it. Or in the alternate scenario, if knowledge is stagnant throughout a time period then it is only our perception of that knowledge that changes or evolves. Similar to language, the means of gaining knowledge change via perception yet the knowledge itself remains the same. Regardless of whether knowledge is stagnant or evolving, we are still stuck in our realm of perception. So contemporary philosophers claim that if the nature of knowledge is to be perceived, then that is the nature of reality. Perception becomes reality, according to today's society of philosophers, because the reality that is outside of perception can never be figured out to be real or not real.

Affirmation of knowledge builds up more knowledge, and this cycle eventually defines knowledge itself. A consensive thought on a certain topic becomes more real over time as the consensus is examined by 'people of authority and 'scientists'. Even though it may be a deception, our brains and our society continue to build upon its previous findings. The axioms built-in mathematics and physics may be flawed, yet since they are the foundation of that realm, they are forever to be true.

So where does this affirmation theory lead up to? It goes back to meditation because essentially meditation relies on the concept of introspection. Zen Buddhists believe that all the knowledge in the world is within us and all that we have to do is be mindful of it. This realm of introspection is the highest way of attaining knowledge, and it doesn't fall into the problem of perception. Nor does it fall into the problem of affirmation theory. To criticize meditation theory is like being blind to the ultimate truth. The problem of language kicks in as soon as you try to explain what meditation really is. But meditation is the means of gaining knowledge that does not fall into perception or affirmation of perception. The only way to determine whether the meditation technique is viable to attaining knowledge is by doing meditation. Up to this day, there hasn't been a monk who has falsified his own belief, which shows that those who meditate already know that their practices cannot be proven wrong in any manner.

Conscious Effort to Attaining Knowledge

All philosophies and theories about the nature of knowledge attempt to get to a concrete result. Yet does this desire to figure out the nature of reality affect the attainment of knowledge. We know our perception predisposes the realm of knowledge, but do our expectations and desires also predispose the quantity and quality of our knowledge?   Many psychologists have referred to this phenomenon as salience effect. Salience is just a conscious effort of attention. The salience effect suggests that we cannot gain all the knowledge there is at the exact point. This is because to know something, we have to reach its depths. And in this process, we are attentive to one topic of knowledge, while not attentive to another topic. This means that while discovering something, you are indirectly not discovering something else. Although one might just say: 'I can figure out one thing at a time', the individual doesn't realize that figuring out two dependent events not simultaneously leads to an incorrect conclusion. Thus the attainment of knowledge itself is paradoxical because to attain knowledge is to examine it individually and separately both simultaneously. How does one do this? We believe that the process of seeking knowledge is different from the process of attaining it. And thus they cannot happen simultaneously. Yet like mentioned previously, the process of attaining knowledge is more likely than not flawed by the problem of language, the problem of perception, the problem of affirmation and so forth. Thus in order to reach complete knowledge without such obstacles, one has to look at attaining knowledge as the same thing as being in the process of attaining that exact knowledge. So how does one learn and be in the process of learning at the same time? The simple answer is meditation.

Meditation acts as a means of attaining knowledge, yet in the process, it gives knowledge itself. As an individual meditates, he or she is in the process of attaining knowledge outside of perception, while simultaneously being in a state of having that knowledge. The concepts of Buddhism, as a philosophy, show up a lot in the attainment of knowledge. It isn't logic, science, or language that produces knowledge, but rather it is the act of being knowledgeable itself that produces knowledge within. Introspection, thus, lies at the bottom of seeking knowledge.

Socrates and The Buddha
"The only true wisdom is in knowing you know nothing" - Socrates.

Nature of Reality

Socrates was an intelligent man, and he denied that he was intelligent. He told others to do the same, as he believed virtue was knowledge, and a virtuous man would never boost his ego. But it is paradoxical to say to know is to not know. Furthermore, Socrates thought he knew that he didn't know when in reality it might have just been a self-fulfilling prophecy that happened to be created for Socrates. Socrates' conclusion is somewhat negative and nihilistic. If there was no meaning in anything because everything fell into the realm of deception then the world wouldn't manifest itself. If everything was for no reason, there wouldn't be life or death for people would just die because there would be no point in life. Such conclusions were not intended to be made by Socrates but happened unconsciously. Socrates wasn't able to find knowledge outside of perception, thus he claimed that no humans would be able to find it either. Yet Socrates' failed to examine his flaws. Socrates knew he had flaws, but that is where he stopped looking. If Socrates had tried to solve these flaws of perception then eventually he would've realized that humans aren't as helpless as he thought them to be. On the other hand, 200 years earlier in 600 BCE, Siddhartha Gautama was able to spread the concepts of Buddhism and the practice of attaining ultimate knowledge: nirvana. The Indian civilization is the longest civilization on the planet of earth. Oftentimes, people neglect the intellectual findings in Eastern Asia, particularly India. The Buddha had realized exactly what Socrates had realized, but only 200 years earlier. Yet Buddha approached the problem of perception differently than how Socrates did. Instead of falling into pessimistic thoughts, the Buddha realized the concept of introspection. With introspection, Buddha was able to realize that the act of gaining knowledge was nothing but an act of introspecting and realizing the inner self's potential. While Socrates never even thought about solving his problem of knowledge. Socrates believed in the helpless nature of men, while Buddha believed in Dharma, which was the right path to living in harmony with the inner self. One can debate about this topic for hours, or one can learn the

Knowledge

lesson from this historic discrepancy. The bottom line is language and people are the means of knowledge, the individual, though unaware, is the knowledge itself.

3. Sense Perception

The five senses

Given that more often than not we live in the realm of perception, it is important to examine how different perceptions work. Language is one way to perceive and categorize knowledge. sense perception is another, more oft-used method to gain knowledge. What is sense perception? Simply put, it is the understanding gained through the senses of touch, sight, sound, smell, and taste. These senses are the foundation of perception.

For a sensation to happen, there must be an external stimulus that starts neural activity in the brain. Neurons are biological cells that carry information through ion signals. When touching something, the sensory neurons in that part will activate(with an action potential), link with other neuron chains to form a synapse, and eventually go to the brain, which will decode the neuron message. Note that the eyes, ears, skin, tongue, and nose aren't the ones perceiving the information. These organs have receptors that collect the information and that information is decoded in the occipital and temporal lobes of our brains. A binary system is used by the brain to fire neurons and decode messages from the peripheral nervous system(which is throughout the body) to the central nervous system(which is in the brain). A binary system is a system in which 1's and 0's are used to code specific information. 1's mean to transmit and 0's mean to inhibit. Sensory Neuroscience is much more complicated than this, but the essence is that the way the brain's nervous system decodes information is different from the way our senses decode information from the external world. For example, a book is a physical object for the eyes, yet for the brain, it is composed of neuron exchange of electrical signals. Famous philosopher, Locke, argued that the brain doesn't decode information to be relative to space and material. But rather the brain produces conceptual ideas that describe the information, which then is perceived in space and material. Our bodies have more than 80 billion neurons and these cells allow for sensation to happen. It is important to understand that sensation and perception are different. Which one happens first? That is a philosophical question that will be examined later.

The only way to physically interact with the world is by touching objects in the world. Physics explains how each object has atoms, protons, neutrons, and electrons and these allow the universe to exist independently without the problem of perception. Objects of touch can be categorized into three different forms: solid, liquid, or gas. Objects that do not fall into these categories are practically not sensible by the skin but are possible to be comprehended by the brain. For something to be sensed by the skin, it has to have a form, but for something to be sensed by the brain all it needs to have is a receiver. Thus perception can happen without the realm of materials and space, but sensation needs a certain physical platform to occur. Hearing, smelling, and tasting occurs in a similar fashion. All sensations need an external stimulus to trigger neural activity. But what if this external stimulus is manipulated to trick our brains? The rubber hand illusion demonstrates how one can believe to sense pain in the brain by deluding their eyes to believe that a fake hand is their actual hand. It would be worthwhile to watch BBC's video on the rubber hand illusion to better understand this phenomenon. Optical and hearing illusions also work to manipulate the stimuli and cause a different response. So if our senses are so easily deceived, what then lies at the core of deception?

Salience and Differentiation

Deception occurs due to salience. Just like salience affects language interpretation, salience affects our sense perception. As humans, we focus our sensations on one thing, and if this focus is distracted, perhaps that sensation isn't even perceived. For example, going through a crowd at a station means pushing and being pushed.

Yet this sensation never seems to cause pain for us because we aren't salient of that pain. Our focus is on the grip of our backpack or the child on our back. There are so many sounds and voices in a supermarket, yet they are never sensed because we pay attention to only that which pertains to our benefit. Thus we predispose our sensations and often receive less information than we could've if we were a bit more alert and not so selfish about being salient to only that which pertains to individual benefit. The rubber hand illusion works because the experimenter is able to change the recipient's vision and touch focus to something different than their regular focus. The recipient knows that the rubber hand is not their own hand, yet they feel pain when the experimenter hits a hammer on the rubber hand. It all happens because the recipient paid attention to the fake hand while receiving the touch sensation on their actual hand. This discrepancy causes them to deceive themselves into perceiving a nonexistent sense. If anything, this experiment suggests that the five senses are interconnected and that people can be conditioned into believing a fake sensation. This process of reconditioning the neuron signal in the brain is called neuroplasticity. Neuroplasticity is used therapeutically to treat people with prosthetic limbs. After enough practice, an individual could program their nervous system to sense without a real arm. The machine's movement and sensation correspond to the individual's perception of his own movement and response to sensation. Thus saliency and conditioning practices allow humans to trick their senses into forming new memories and the new perceptions of sensations.

    Differentiation is similar to saliency but it gets to the point of how sensations cannot be formed by themselves. In order to form a perception of sensation, there must be a pre sensation and post sensation. If an individual has a sensation of being really cold, and if that individual enters a warm bathtub, then that person would feel really hot, if not feel as if burning. This is due to homeostasis and the fact that their pre sensation was completely different from their post sensation. Sensation follows the rule of time thus what happens before and after is really important. Frostbites can occur because the body temperature goes from really cold to hot as an individual washes their cold feet with warm water. It isn't that the water itself is burning hot, but it is rather the perception of that sensation that makes the water warmer than usual. The smell or a taste of a nice meal overrules the post smell of the environment or the post taste of the overly sweet dessert. Eating sweets before drinking milk leads to perceiving the sensation of milk to be not as sweet as the regular milk. Thus differentiation in pre and post sensations creates extreme perceptions that serve to be there for temporary moments of time until the body reaches homeostasis when it goes back to regular mode. All these examples serve to show that sensation is dependent on previous sensations. In fact, this is how memories of sensations can cause the perception of those sensations to be stronger than regular. A hot cup of coffee with a loved one carries a strong feeling that then brings joy every time you drink a hot cup of coffee, whether that be with or without the partner. And if that partner were to break up with you, the exact same hot cup of coffee seems like the worst thing ever.

Thus sensations are manipulated by differentiation and saliency. Each of these factors allows the perception of sensation to change and even produces a strong memory of that false sensation to be remembered as a way of bringing pleasure next time the same false sensation occurs. At last, sensations are interconnected and pleasure in one sense can lead to pleasure in another sense just like the pain in one sense can lead to pain in another, ultimately leading to pleasure or pain in the brain. Nevertheless, it isn't the eye that hurts or the nose that hurts. Rather it is the brain that produces that perception of pain in that area.

Sense Identification

Earlier on, there was a question about whether it is sensations or perceptions that occur first. It makes sense to say that sensations cause perceptions right? Because if pre sensations result in a different perception then that means sensations come before perception. However, this is where the realm of paradox begins. Is it that we perceive sensations and then cause them to align with that perception after the actual sensation, or is that we sense something and then perceive it to be. Is an object identified before it is even sensed? That is like saying the universe already knows what is going to happen but it is only aware of that when the event actually happens. Certainly not, one might think, right? However, the latest neuroscience studies show that the brain already knows the sensation before it occurs. The concept of efference copy is that the signal that is sent from the sensory neurons to the motor neurons is copied and sent to the cerebellum which is the 'thing that approves the message.' The action of the muscle movement is anticipated even before it occurs by the cerebellum, and then that action is executed by our muscles. In other words, we perceive a certain action based on earlier memories of performing that action, and in this process, we anticipate a sensation, that is then performed by the muscles. If the anticipation is incorrect, then new memories are formed to correct the anticipation for the future.

    This concept of anticipation of sensation was discovered way before the scientific field of neuroscience started to grow. In many Asian cultures, saints were able to predict the future of kings and queens. Similarly, monks were able to predict and control their thoughts before they arrived at their minds. The so-called 'supernatural' powers these monks and saints had are being examined just recently and slowly have been accepted by the scientific field. Surely these monks didn't know a thing about neuroscience, or did they? Often times Buddhists are seen as hermits who close their eyes and fall into unconsciousness. Yet, there is a good chance that such scientific knowledge was self-learned by the monks way before it was ever able to experience or sensed by the eyes. The concept of neuroplasticity is produced by electrical signals in today's day and age. Yet The brain can be changed via experience, as the rubber hand illusion showcased. We don't need big machines and computers to change our own thoughts and brain. It comes down to the same concept: Actions of the mind produce actions of the body, thus controlling the actions of the mind(or brain in the scientific context) can lead to controlling the actions of the body. Medication can be the uttermost perfect way of reprogramming the mind, which is the same thing as going through a neuroplastic surgery that costs millions of dollars worth of equipment.

Preference and Equanimity

Pain and pleasure are sensations felt on a daily basis. Not just physical pain, but mental and emotional pain is also a sensation in our minds. How does the sensation of pain relate to our perception of them? Pain and pleasure are subjective, so each individual builds different stimuli and responses to a particular event. One might find pleasure in walking out in the snow and appreciating the beauty of nature, while someone else would hate the experience of walking out in the snow because of the low temperature and slippery ground. In the end pleasure and pain derive from preference over something. Some prefer the sensation of cold, while others don't. This leads to each of them feeling pain or pleasure. This preference is built up over time and objects of desire seem more likable and objects of aversion seem more dislikeable. This difference accentuates as the individual expects to always receive pleasure. The fundamental concept of humanity is that we want more pleasure and less pain. In this process, we condition ourselves to like some things and dislike other things.

Preference for objects happens randomly, yet we make sense of them as if those preferences were destined for us. A baby has no preference over what food to eat, yet it just so happens that as the baby grows it prefers some food over the other. Nobody is predisposed to feel pleasure from pain. Pleasure and pain are products of preference. Thus those who think their lives will always be depressing and lonely are actually the ones who prefer that loneliness or are the creators of that preference. Thus self-fulfilling prophecies and wishful thinking are key concepts to living a healthy life. Unfortunately, there are some people who think very pessimistically about life.

Nietzsche's teachings about nihilism are very ignorant of their own flaws. Nietzche believes that life is full of suffering, and there is no way out of it. Nietzche's pessimism arrives from the fact that there is no good in life. In fact, the Buddha had realized this way before Nietzche's time period. Just like Nietzche, Buddha also believed that humankind is prone to experience pain. Yet he believed that pain is a production of our own mind. Thus one can control the process of feeling pain by controlling the process of preference. Since humans prefer one thing over the other, they are attached to their preferences and thus they get agitated or feel pain when those preferences are not given to them. Yet, nature doesn't always listen to each and everyone's preferences. Nobody gets exactly what they want at all times. Thus it is true that we are stuck in a cycle of feeling pain. However, the pain is a cause of preferring one thing over the other. In fact, having no preference leads to a less painful life because the individual is accepting of all stimuli, regardless of their response.

In the latest studies of neuroscience, scientists have claimed that there is no such thing as pain. Nor there is such a thing as pleasure. There are no events, or stimuli strictly associated with either pain or pleasure. It is rather the perception of pain and pleasure that brings pain or pleasure to the mind. Our brains adapt to painful sensations and we perceive them to be less painful later on. PTSD or Post Traumatic Stress Disorder is often resolved by stimulating a person's beta-blockers which are meant to block adrenergic receptors responsible for the person to feel pain. Instead of feeling pain, the patient is forced into pleasurable stimuli that overrule the traumatic emotion. This way, the specific neuron communication of pain soon transforms into the communication of a nonpainful event. However, neuroscientists can solve PTD only temporarily. Similarly, psychologists can only temporarily help depressed individuals. To get rid of the root problem, the individual has to control their mind and not be in the trap of constantly looking for preferred objects. Because the preferred objects are the exact reason for the individual's experience of suffering and pain.

    Monks live in forests without preferring to eat a certain food, sleep on a certain platform, or even look in a certain manner. These monks have little to no preference about what they want. Yet they are the happiest men in the whole world. By controlling their minds, the monks have conditioned their brains to align pain and pleasure. This allows them to never perceive pain or pleasure. This concept of equanimity is one of the most important lessons of the Buddha. If a man were to live life equanimously, he would be happier than ever. Pain and suffering are inevitable for an individual who sees preference in the materialistic world. Nietzsche and Socrates were close to discovering the nature of reality. Yet both of them failed to realize the power of their own minds. The mind creates reality and emotion. The mind has the greatest force on the individual. Thus once again, introspection and mindfulness serve as key techniques for escaping the realm of inevitable pain and suffering and reaching an equanimous and calm state of mind.

4. Concept of Education

Teacher and the Student

In seeking knowledge, one must acquire it from someone, or something. The teacher and the student are often seen as different entities when it comes to education. Teaching cannot occur without a student, nor can learning happen without a teacher. The purpose of education is not to gain more and more knowledge, but to understand the essence of gaining knowledge. Once we understand how to gain knowledge, then we can gain any knowledge there is in the world. Education systems are often good at getting information across to the students. They do this by incentivizing students with a grade mark or a standardized test that boosts their ego and makes them study more. Yet this is a flaw in the education system. Proper education is teaching students how to teach themselves. Education and success are often related to one another. Yet the concept of success is not determined by the education you receive, but by the education, you provide to others. The fundamental concept of education is to self-learn and guide others to independent learning as well.

The teacher not only provides knowledge about a certain topic or phenomenon but also learns in the process of teaching. Similarly, the student is not only responsible for gaining knowledge but also for reconveying this knowledge to themselves and others. In this way, education is maximized and knowledge is spread to as many people as possible. In reality, teachers and students do the exact same thing, yet we view them separately. Respect and attention to teachers require them to be on a higher platform. Thus society perceives them as better than students. Education systems must equalize students and teachers to allow teachers to learn from students during the education process. In a perfect world, everyone is their own teacher, yet this requires a lot of focus and self-determination. Thus, an easier way to teach is by learning how to teach ourselves before learning any concept.

If the information exchange is happening only one way, soon prejudices and deceptions will occur in education. Teachers will always perceive that their knowledge, passed down to them by their educators, is the ultimate knowledge that must be. Conversely, with the process outlined above, teachers value the knowledge they received rather than a conception of absolute knowledge. Thus, education is nothing but a circular process of teaching others to teach themselves. The teacher learns how to teach themselves, by becoming a student, and the student learns how to teach others to learn on their own, by becoming a teacher.

Behaviorism and Predisposition

All teachers and students are infants at once, and they are predisposed to certain norms set by their families and environment. Behaviorism at its core explains the natural process of learning in a child that happens due to the child's environment. Thus the child is fully dependent on its family. We grow up to learn one language and become part of one culture or religion. This initial state of learning influences later parts of our education process.

Infants observe their parents and thus they learn to do exactly what their parents do. It is up to the parents to mold the child into a good human and pass down values that allow the child to develop and have its greatest potential. Yet the current era, being full of gadgets and technology, already predispose infants and children to be dependent on exterior means of knowledge rather than their own interior processes of the mind. When growing up, a child often sees their parents working on their computers, a sibling playing a video game, or even a grandmother watching her favorite television show on the couch. Without any thought, the infant predisposes itself to believe that the world revolves around technology and computers. This inhibits the infant from even thinking about the world that is inside its own head. Technology is just one means of life in today's day and age. This generation is consumed by such a vast materialistic world and such materials of desire often lead the child into believing that the physical world is far more necessary than the interior world. The infant doesn't decide what he can condition himself to be at such a young age. All the infant can do is the condition itself according to its surroundings. Thus these infants grow to produce more infants of the same type, and we have a world full of people who have been predisposed to thinking that the world is nothing but a realm of physical objects.

B.F Skinner's classical conditioning theory is worth noting. Classical conditioning happens when a child associates stimuli with a specific response. If the same stimulus results in the exact same response over and over, the infant conditions itself into believing that there is no other response to that specific stimulus. For example, if a mom rubs the child's back every time before feeding milk to the baby, then the baby will create a psychological pattern that will allow it to realize that rubbing on its back means it is time to be fed. This classical conditioning happens with every pattern the baby senses. If the baby sees its mother on her phone, laughing, then the baby associates that laughing face with being on the phone. Yet babies can also be dehabituated, which is a process of changing the correlation between certain stimuli and their response.

Dishabituation happens when an earlier learned stimulus produces a different result over and over again. This leads the baby to produce different psychological patterns about an event and its results. If anything, classical conditioning shows how easy it is to condition a baby to act or behave in a certain way. Classical conditioning occurs in adults too, yet they are more capable of dehabituating anything that seems immoral or unexpected. However, babies haven't gained consciousness of themselves and morality, thus predispositions are crucial for a baby and its upbringing. As they say, a lie told enough times becomes a truth. If generations upon generations raise kids in a way that enforces kids to learn the wrong thing, then the world becomes full of people that have deceptions and predisposed lies about reality. Then again, is today's world any different?

Concept of Education

Humans are social creatures, and without a doubt, humans are prone to conformity based on society and the environment. This is a psychological fact. Yet knowing this, people continue to display the wrong image of themselves in front of the kids. There are two sides to educating kids in the right way. One is that you show them a world that doesn't require human creations such as objects of the materialistic world. Rather, you instill a philosophical realm in the child. This doesn't mean you read The Republic to a 3 months old child, but somehow show the child a natural way of life. That is by talking to the child on long road trips, in which the child will be able to look outside the window. Or even just visiting different architecture and cultural places. Obviously, a child younger than 3 years has no cognition of sensation, and most of the time, the child is asleep. But after 3 years old, the child learns to develop a perception of the world. And this perception is deep-rooted in the child's psyche, so it is important to allow the child to do something other than watch television or play video games. The second way of education is by setting an example. The child will condition itself according to what the parents put forth. Thus setting an example that doesn't distract the child from thinking about reality is important. Of course, teaching morality is also important because you don't want to raise a child to be arrogant and selfish. There are a lot of debates about whether humans are naturally selfish or not, but it all depends on how they are raised. No child will do harm to another child for the sake of its own pleasure. That impulse of hurting someone else is only viable if the child receives a poor environment at home. I am not a developmental psychologist, so I can only go so far about the nature of development. Yet the bottom line is habitualization in babies happens very rapidly thus it is important to provide the baby with right knowledge about morality and about a realm outside of the physical world.

Institutions of Education

Now that we have examined the infancy period, let us move on to youth and teenager life. It is true that certain things are predisposed upon us during infancy. However, we learn quite a bit about the basics of life and knowledge during the first 18 years of our life. School is a place where children go to 'learn' new things and to make new friends. It is important to have socialization in order to be part of this society. In order to contribute to society, one has to be able to connect and communicate with other fellow humans. School is a perfect way to allow this communication to happen naturally.

Yet, there are about 4% of students who are homeschooled. These students are nonetheless as smart if not smarter than those who attend school. Schooling, whether it be homeschooling, private schooling, or public schooling, is compulsory around the world, exceptions being areas influenced by orthodox religion or monarchy. Education is regarded as an important investment across all nations, regardless of their political or economical status. So what is it that makes education so important? It can be argued that education is required for people's safety. An educated man would think twice before acting on impulse. Yet morality isn't something to be taught, rather is something to be experienced and self learned. So then why do schools exist?

Schools exist to make people competent enough to contribute to society. Whether that contribution is through knowledge or labor is up to the individual's capacity. The term blue-collar job is referred to those who do labor, while white collared people are those who have the job of passing down certain knowledge. Education promotes people to become white-collared citizens of society. Yet in this process education systems forget to teach people how to do labor. Labor is seen as something nonintelligent people do. Yet any intelligent person would know that a balance between labor and contribution of knowledge is key to life. We rely on the bus drivers, snow cleaners, grass mowers, and other individuals that do basic labor for us. These people are necessary, but we allocate only certain people to do this job. Yet there is a school of thought that states that if everyone were to do their labor, that is things such as picking up their trash, cleaning their offices, making and growing food, and all other necessary labor, then nobody would be necessary to do the so-called 'blue-collared job.' If everyone knew how their appliances work, and how houses are made, there would be no need for carpenters or builders. In this way, everyone would know exactly how to do their labor without harming anyone else. There are some political issues with this process, but that will be examined later. The bottom line is that if schools were able to teach people how to be both white and blue-collar citizens, then everyone would be able to contribute to the knowledge systems. When self-sufficiency can happen in society, education of knowledge systems will be better implemented. More people will be able to spend their lives contributing to seeking and teaching knowledge.

    Schools do a good job of teaching the natural sciences, mathematics, writing, and the humanities. Yet philosophy and cultural studies are not mentioned until college. Philosophy and psychology classes are not even mandatory at most high schools. Philosophy is the study of life, and if people can't study philosophy then what is the purpose of living? That might seem extreme, but philosophy is necessary for humankind. It is through the philosophy that people will be critical thinkers and go out of the box to cultivate knowledge not known to the current society. It isn't that schools should stop teaching students the other subjects, but the argument is that all schools should instill some type of class that invokes students to think beyond the realm of perception and the material world. To teach is to allow students to teach themselves. And it is the philosophy and internal thinking that will invoke students to think and learn on their own.

    The word philosophy means love of wisdom, Thus by invoking a sense of love for knowledge, teachers will be able to teach students how to learn and love other subjects taught in school. There are some organizations such as the International Baccalaureate, that have a curriculum involving classes such as Theory of Knowledge or epistemology and assignments such as Prescribed Title, that allow students to make their philosophical arguments about reality and perception. This curriculum should be accessible to more schools to produce men and women of high intellect. Many western and eastern philosophers have aided us to the point where we can take their works of knowledge and use them daily. Theory of knowledge is nothing but practicing knowledge in daily life through multiple perspectives and ways of knowing. Similarly, philosophy isn't just thought put into words, but rather words that are meant to invoke thought. And it is only through such thoughts that humans will be able to one, escape their limitations of thought and action, and two, go beyond the realm of perception.

Examination of Knowledge

Education institutions measure the success of education level by standardized tests and GPA, or grade point average. These methods are somehow communicative of a person's intelligence level and the school's education success. These measurements give students an incentive to work hard and be able to graduate with a good GPA and a good standardized score. Both of these methods are comparative methods which means they scale individuals based on their rank in their grade group. Ranking individuals allows the individuals on the top to feel like they are smart, thus to keep that recognition, they continue to study and work hard. On the other hand, those who fail and those who do not rank on top feel shameful thus they believe that they can never succeed in life. This system of grading and ranking is very paradoxical because it attempts to improvise student's motivation and allow them to compete with one another at a friendly level. Yet in reality, the grading system and tests only create an inferiority complex among students. Those students who have access to paid classes, technology and smart enough parents, are able to score high on their tests. While those students who cannot prioritize education first and have to support their family either due to economical struggles, or coping with aggressive family members, will score low on their tests. Thus teaching students how to handle their psychological and mental state is far more important than asking them to memorize a math theorem. No student is a student unless they are able to study inside and outside of a classroom. If a student goes to school only as a requirement of the law, then that individual is simply wasting their time. Education is not something that happens when you go somewhere and do something only in that place. Rather it is a process of developing commitment and sheer motivation to grow as an individual in and outside of the school grounds.

Any intelligent individual would know that no sort of examination or recognition is necessary to prove their intelligence. In a pure education system, examination won't exist as means of incentive, but as a means of true self-reflection. Students emphasize the grades they receive, never understanding what is it that they missed or got wrong on their exams. The process of growing by learning and self-reflecting is the purpose of school and education institutions. If education institutions do not pay attention to their student's will to study or their student's psychological traumas, then those students will never appreciate the process of learning, and in return, teachers will lose their purpose. To make students love learning, the institutions and teachers must first solve the problem of mental health and improve students' will to learn. And this must be done not by forcing them to study but by invoking self-reflection in a way that allows the student to improve for their good.

Paradox of Technology

One of the major goals of today's society is to develop computer systems to make several processes efficient. We no longer have to sit next to the fire because we have the heater. We no longer have to walk hundreds of miles to visit family because we have public transportation and video and audio calls. As society advanced, electricity and technology became major goals of improvement.

Nature of Reality

Everything revolves around making things convenient for humans. One might say that human ingenuity is necessary for humans to survive and evolve if that is even a thing. Can humans evolve any further? Nonetheless, the STEM career fields are the most popular in the world. All countries want to produce technological products that surpass their earlier models. In this process, humanity has lost its pure essence of life. While constantly looking to become faster, efficient, and productive, humans have forgotten their true simplistic nature.

The development of technology is a never-ending process. Humans will always find something to tweak or something to make better to fulfill their tasks. Complexity is more valued in today's day and age than simplicity. We have become ambitious in thinking about the future and the next generations. The process of AI creation and autonomous machines has already started to come to life. NASA is trying to discover new planets for life. Nuclear physicists are trying to come up with better and more destructive weapons. Companies like Google and Amazon have been developing their software to such an extent that it tricks consumers into buying products and living in the cycle of wanting more. Computers are becoming more adapted to the examination of body parts. We have technologies to measure heart rate, brain activity, and other complex phenomena that we wouldn't be able to measure without such advances in technology. Yet all of this is simply a never-ending process. Not just that, technology will only get worse in the upcoming era. My theory is that technology will instill laziness in individuals because robots will take over all the labor jobs. Individuals will be able to spend more time trying to figure out how to work out a robot than they will spend time trying to figure out their selves. External materials will condition our brains to find pleasure and pain based on external stimuli. Soon the human kind will forget the notion of peace because everyone will try to rise upon one another. This competition for technological advancements is already in the process. Countries are banning certain technologies in certain areas. Apple and Samsung are fighting one another to produce better products. Yet in the process, they are corrupting society. All technological industries are indirectly corrupting their consumers. Consumers are deceived into thinking that technological equipment will make them happier or allow them to live efficiently. Yet instead of making consumers happy, technology is only creating problems for them. And the fact that there will always be better products in the future just creates a never-ending cycle that disables people from introspecting. This isn't supposed to be a rant of any sort, yet the message is that technologies are not simplifying the lives of humans, rather they are making them more complicated and troublesome.

What would happen if we stopped our technological advancement? What would happen if people didn't have access to the internet? What would happen if there was no public transport? To be quite frank, absolutely nothing. Life would be better than ever. We would clean with broomsticks instead of vacuum cleaners. We would walk to the local store instead of driving. We would have to physically write letters instead of sending emails. The trivial tradeoffs seem nothing compared to the benefits such as less stressful life, more family time, more introspection, and at last the ultimate ability to escape the materialistic world. Simplicity and minimalism are seen as predated concepts in today's era. Yet in practicality, those two concepts are the key to happiness in life. Making life easier doesn't mean it will be happier. In the end, technologies are nowhere as complicated as humans themselves. Before trying to study machines, humans have to study their selves. Thus humans will soon realize that instead of making their materialistic lives efficient, their purpose all along was to make their mental lives more peaceful and simple.

The paradox of technologies is still debatable because there are so many companies involved in technological production. Shutting all these down will lead to a very hard decline in the economy and soon to inflation. Although the economy would fall, I believe that over time, humans will be able to cope up and learn from their ancestors. Maybe that's where the theory about the universe collapsing comes from. One day, humans will be so advanced and so superior that they will be bored of themselves. One day, technologies and robots will be so advanced that humans will have no purpose for themselves. It is only then that they will realize that technological advancement was never humanity's goal. In its process, nuclear weapons and other harmful productions only caused the loss of many lives. All along, humans had a choice of picking simplicity and satisfaction or ambition and unnecessary advancement. But humans chose to be greedy and feed their materialistic desires. It is only then that they will realize that what they thought they were was different from what they truly are.

The paradox of all other career fields

Technologies will affect other fields too such as health care communities and law systems. One could easily make an argument about how we need technologies to save people. Yet a simple solution to that would be to be meditation. Physical injuries are caused due to not being aware such as being drunk while driving, tripping over a fallen branch, or eating too much food. These problems will solve themselves when humanity will realize the potential within them. Crimes will happen less frequently and society will put more emphasis on developing non-technological ways of producing medicine. We don't need to get rid of all technologies. Although containing the usage and production of technologies will help aid individuals to a more peaceful life. Doctors won't even exist in a system that teaches everyone to maintain their health. Each individual would be responsible for keeping themselves healthy, and if they were to be sick or diseased, family members would be willing to help them at no cost. Such an independent system would allow individuals to be more alert and attentive to themselves, instead of being careless about their bodies.

Like all theories, my theory is only a possibility. I can't tell if technologies and health care systems will develop in a certain way in the future. But I truly believe that humans will be healthier and happier if they started to discover knowledge about themselves through the means of simplicity and minimalism rather than technological and materialistic advancements.

Utopia on Earth?

The theory about reducing technologies, and increasing simplistic and minimalistic ways of producing and teaching knowledge is a concept of utopia. I can't be naive enough to believe that humankind is capable of producing such a system that benefits everyone at all times. If it was so easy, humans would've done it all along. Yet the unfortunate fact is that humans are often selfish, unconsciously. Whether that is due to a predisposition toward free will remains a mystery. But one cannot wait for a society to change itself. If an individual becomes helpless realizing that we are doomed all along then that individual is more likely to give up and never try. If each individual tries to live a minimalist life that invokes more self-reflection and thinking rather than consumption of external products, then one day we could truly have a perfect utopia.

Utopia or a perfect world is attainable on this earth. Utopia isn't achieved by an economical promise of land or a perfect societal system that produces morality. Utopia happens when people can educate themselves in the right way, and lead others on the right path of introspection and minimalism. Not only will all specimens be able to educate one another, but they will also be able to make one another happy with no intention of expecting anything in return. Wealth and reputation would no longer be barriers to happiness because everyone would help one another reach Nirvana. Nirvana isn't a spiritual or religious process, but rather a built-in way of reaching peace and tranquility. In Nirvana, the individual seeks for nothing, knowing that they already have everything they ever need.

5. Concept of Science

Realm of Perception

Sense perception shows up in everything including science. The only way to understand the nature of the physical world is by using the natural senses given to us. In the realm of perception, one has to use sense perception to gain knowledge. What lies outside of perception needs other methods, yet if an individual wants to gain only that knowledge which is in the realm of perception, then sense perception and other studies of science is good enough.

Technology as a tool to examine a phenomenon in the world. Last chapter, I said that technology was not needed to make us happy, yet that isn't entirely true. One cannot completely get rid of technologies and all the efficient ways of gaining knowledge just for the sake of simplicity. The realm of science and perception heavily requires technological advancements and that is needed for this realm. Technologies can be a good thing if they are used for the understanding of our world, the physical world. Thus there has to be a balance between simplicity and ambitious advancement.

What is science? I would define science as the knowledge about the physical world gained through experimentation and repetition. When first talking about knowledge, science was heavily criticized due to its incompatibility with other ways of knowing such as faith and emotion. It is hard to define what knowledge truly is, but we can categorize knowledge by the realm it is in. Science happens to be in the realm of perception. There are two worlds we live in: one is the conscious world, and one is the unconscious world. And science examines the conscious world that we live in. The theory about introspection and meditation has nothing to do when it comes to conscious knowledge and understanding of the physical world. It is only through science, that we can understand what the physical world is truly about. Although the metaphysical theory mentioned earlier contradicts the concept of there being a world separate from our minds, it is important to have an open mind towards all fields including science. Materialism, though criticized, helps many scientists learn about the world around them. Whether that world is a creation of their mind or if it be seperate from their mind is something that can only be answered by escaping the realm of perception. Yet most people remain in the realm of perception, so it is worthwhile to examine what it is that scientists truly do and how it is that they come to theories about the physical world.

Experimentation and Reasoning

Science relies on two key ways of gaining knowledge: experimentation and reasoning. It is through these two methods that scientists are able to conclude or come to an answer about the physical world. Experimentation is the process of examining how something affects something else. Obviously, that is a third-grade definition, but it boils down to the concept of experimentation. There is an independent variable, which is the thing that is changing, and there is a dependent variable, which is the thing that is being changed. Alongside this, there are control variables that are meant to be kept constant throughout the experiment. Some experiments examine whether the independent variable correlates to the dependent variable, while other experiments examine whether there is any causation between the two. But even a third-grader knows that causation is no correlation. This foundation of variables and concept of causation allows scientists to see what factors affect or correlate to our physical world. Experimentation is very useful, but it could trap people into believing that correlation is causation. Just like conditioning, if a stimulus or a variable always correlates to another stimuli or variable, our minds easily change correlation to causation. Just because the earth didn't collapse until today does not mean it won't collapse tomorrow, yet scientifically that is nearly impossible because we believe that the earth is still in its infancy period and that there are billions of years left before the earth collapses. But it seems ironic how we create patterns about what will happen based on what has happened. In experimentation, scientists have previous data that helps them hypothesize about the experiment. If the previous data or previous knowledge conflicts or completely opposes knowledge that is produced currently, then scientists reexamine their experiment or try to figure out another way of making sure everything is complementary to previous knowledge. Many scientific theories have failed, and many have been successful. Yet, something is successful in the scientific field only because it has not been proven wrong by anyone else. As soon as a phenomenon is opposed or is being proven incorrect, then that phenomenon becomes incorrect. Thus through time, science has improved and the truth about the physical world has 'changed' for the public. Regardless of its success, science can only describe how things happen relative to our perception and time. Hundreds of years earlier, scientists believed their theory about how the world worked was true, yet now most of that knowledge is proven wrong. Similarly hundreds of years after us, there will be new types of scientists who will reject earlier scientific knowledge. All lives are part of history, and as time progresses so will the knowledge about the physical world. But the method of experimentation will remain flawed for the entirety of its existence. Experimentation will always be dependent on technology and the ability of individuals to not be biased. Thus truth in the realm of science is a combination of the truth out there, and the truth that the individual makes by his or her mind. Thus it is perfectly fine to continue using science to help advance our knowledge of truth h. Yet this truth is different from the ultimate truth, which is not influenced by our thinking or mind.

    The other methods scientists have used is inductive and deductive reasoning. Inductive reasoning is basically the process of experimentation and coming up with conclusions based on such experiments. It is basically the process of seeking information to conclude a result. On the other hand deductive reasoning is the process of seeking other results that are presented, and using that to extrapolate and come up with one's own conclusions. Scientists use both of these reasoning methods to understand the truth in their realm. Inductive reasoning is used less often than deductive reasoning in today's age because scientists have trust in previous findings. Deduction allows scientists to build upon previous findings. In this way, information is built upon, and passed down. It is almost like a puzzle. Each piece allows the puzzle to be formed faster because each piece contributes to finding its surrounding pieces. Deductive reasoning follows from the premise that the pieces that are being deducted are actually true. All physicists have built upon one another, yet some have questioned previous predispositions. For example, quantum physics explained in the early 1900's is way different than the quantum physics described in the 21st century. Yet the process of failing and discovering allows scientists to reach closer to truth, if there seems to be one. Thus positive skepticism is necessary for the progression of science. Yet once again, science is heavily dependent on experimentation and deduction so if one phenomenon turns out to be incorrect or misunderstood, then all knowledge proceeding that misconception becomes flawed too. A lie told enough times becomes the truth. Thus reasoning can fall in the trap of producing information that is true only because it is compatible with previous information rather than the truth being true because it was, and will always be true. But truth in the realm of science and perception relies on the observer and the physical entity of observation, thus science can still portray 'untruth' that is conceived as 'truth' by the whole community.

    What makes science unique is that any theory that cannot be falsified is not scientifically correct. The scientific community realizes the human's helpless and imperfect nature. Thus any theory that seems too perfect or seems unflawed is automatically incorrect. This might seem like predisposing certain knowledge, yet that is how scientists deal with uncertainty and flaw. Errors in systems actually make scientists happy because they are then able to make their discoveries one step closer to perfection. Knowing that knowledge is based on the observer, scientists don't think too much about the processes of our minds. In other words scientists neglect the problem of perception by setting restrictions on human capacity. They are more interested about the knowledge that the human is able to perceive rather than the knowledge the human is not able to perceive.

Paradox of Data

Data or information on previous knowledge systems is crucially important for deductive reasoning. There are two types of data: one that is produced by observation, and the other that is produced by patternation and deduction and extrapolation of observation. Most scientists will collect their data, which they justify to be true as per their senses such as vision and touch. This data, which is produced by observation has quite a bit of uncertainty. There is uncertainty from human error; then there is uncertainty from the equipment used; furthermore, there is uncertainty from the observer themselves, as their state of emotion and mind can influence the data collection. All these uncertainties contribute to errors in the result, thus scientific knowledge's accuracy is determined by how low the uncertainty is. Yet, there will always be uncertainty, so long there is an observer. Thus this paradox leads to scientists neglecting human error because it is inevitable in seeking knowledge.

Data interpretation is another skeptical concept. Each scientist examines data differently. Some scientists only conclude results that are true 100% of the time. While other scientists conclude their results when they are true more than 50% of the time. When the public sees this data, they instantly think that if the scientific community agrees with a certain idea, then it must be true. The public overestimates the scientific community as if they are non humans. All humans have preferences and scientists themselves prefer to interpret data in a certain manner. On the other hand, scientists also do a good job of reducing their bias by repeating many trials and reducing uncertainties. The scientific community has expressed a concern about ethics and philosophy as well. Science is only going to advance in the future as it becomes open to other ways of knowing such as religion and philosophy.

Math and Number Systems

Math is the foundation of all sciences. Even philosophy was inspired by mathematics. Mathematics is similar to language, as it describes or communicates concepts of reality through numbers. The numbers don't communicate words, but they communicate descriptions of something. For example, having 2 candies compared to having 100 candies is a completely different experience. the numbers are a way of quantifying objects. Without quantifying things, we wouldn't be able to understand the difference between less and more, big and small, heavy and light, and so forth. Or would we? Math is a way of describing reality. Similar to language, math is the means of gaining knowledge, and not the knowledge itself. Oftentimes, mathematicians realize how their mathematical equations perfectly suit real-life events, and so they believe that math is the perfect way of predicting natural events. But then the question becomes was math invented or discovered? If math was created to better understand the world, it would be similar to language. We know for a fact that language is a social construct. But math is a little different. If we look at math as a concept, rather than the specific laws of nature, then it can be said that humans are born with mathematical skills. An infant knows the difference between big and small or has more or less without knowing anything about addition or subtraction. There have been many psychological experiments that showcase infants' abilities to count and realize the concept of addition and subtraction even before they have learned in the world. This mostly happens because nature follows a set of mathematical laws. Thus the baby does not have to learn math in a specific, rather it can look at nature and its surroundings to understand the fundamental concepts of math.

Math, like all other sciences, has its axioms. An axiom is set up to establish rules in a specific subject. Any science has to have axioms to build upon its knowledge. This then gives way to deductive reasoning. All math relies on the basic operations of addition, subtraction, multiplication, and division. These foundations are the basic concepts of algebra, geometry, probability, statistics, calculus, and even higher-level math. We all know 2+2 is 4. We also know -2 +2 is 0. Addition and multiplication have their counterparts go backward. Yet why exactly does 2+2 equal 4? Why do things add when they are alike, but subtract when they are opposite? This concept is very hard to conceptualize, but the bottom line is that nature that is how things work. There is no need for logic or reasoning to take place. Things are the way they are naturally supposed to be. Yet math becomes complicated when humans try to extrapolate natural and applied math to further level math.

People claim that time is the fourth dimension in the world, and we can come up with equations beyond our senses to describe particles moving in a three dimensional world. But this is all based on deduction and axiom extrapolation. When we extrapolate or take a concept and apply it to other phenomenons, we patternize ourselves and trap ourselves in our own deceptions. No polynomial is a perfect model for any graph, no probability is perfect for any situation, no rate of change will ever be truly examined. Not to be negative, but there is no way math can account for all the variables in the natural world. Nature happens randomly and math cannot account for all random events. Regardless of its use, math needs to be examined further. Math lies between the realm of consciousness and unconsciousness. Because math does not have direct meaning to itself, rather it gives context to the meaning of something else. The binary system of 0s and 1s is a perfect example of how such rudimentary concepts of math can create robots and machines. Mathematical concepts can describe something that language cannot, so math deserves some attention in that field. Yet once again, math can trap individuals into illusionary patternation of equations and data which have their own errors.

Natural and Human Sciences

Natural sciences and human sciences are the other types of sciences that are used in our society. Each one of these has their own methods. But they both go through experimentation and induction and deduction. Natural sciences deal with pure nature, or so they proclaim. While the human sciences deal with humans. Combination of the both gets you in medical and health care fields. Believe it or not, science saves people's lives. But once again, the medical field can only save bodies and not our minds. Anesthetics, vaccines, and all other pharmaceutical drugs work to help our body, but we do not know why they work. Similarly, we can learn psychology and sociology, but never get to the concrete idea of what a human is, how behavior shapes each individual. Science can only get so far as data and experiments can go. But that keeps life uncertain. Each pregnancy has a chance to fail, each disease has a chance of producing more diseases, each drug has its own side effects. This theme of uncertainty is everything in the natural sciences, and more in the human sciences. If things were to happen naturally, the human need not interfere in those natural processes. By doing this, one may be able to increase life expectancy, but they may also be unbalancing the cyclical nature of life and death.

Are doctors really necessary in this world? What would happen if there were no doctors? What would happen if everyone was their own clinic and treated themselves and their families? One might think that people would abuse drugs and medicine. Some would abuse their freedom and hurt others instead of helping them. And so logically, it makes sense to trust only a small number of people who have done their medical school and are purely ethical human beings. So it can be concluded that having no doctors is not a good idea. I think this is a reasonable argument. Until humans are able to live in harmony with nature, they will always be sick and get diseased, so there must be some doctors needed in the world. Yet, if humans abide by nature's law of life and death, and reduced pollution and increased cleanliness, then there would be a point when humans wouldn't need doctors. Humans would live happily without any fear of disease or pain. But this is clearly not going to happen in the future, because as technology is growing, we are using more natural resources and going completely against the natural cycle of life. I can learn nature just by following its laws. Nature asks nothing but sustainability. If humans can focus on sustaining the environment, rather than themselves, then they will inevitably solve their own health problems. Healthy air means a healthy body and a healthy body means no need for doctors. But to say that doctors are useless is a very poor statement. In the end doctors themselves are humans, and are prone to be victims of nature's disaster. Natural utopia goes hand in hand with educational utopia because all humans are able to sustain their own health and the health of the environment.

Human sciences are a little different than natural sciences as they examine ourselves. Human sciences allow us to examine our mechanisms of living. Natural scientists argue that humans are part of nature thus they must be examined biologically. Yet there is something unique about humans that cannot be examined by natural scientists and that is emotions. Emotions give rise to behavior, which is the foundation of many human sciences. Why do humans behave the way they do? What makes a human different from other animals? These questions can be examined through natural sciences, yet they are more approachable in the human sciences. Certainly, an EEG or an MRI scan can tell a lot about a human's behavior. But the same can be done by vocal therapy or psychoanalysis. But I think it is quite foolish to use external tools to examine something deep-rooted internal. Psychiatry and therapy can help an individual recover for a short period of time. It gets rid of the problem temporarily, but the root cause of the problem remains. The root problem of any mental disorder lies in the inability of the individual to deeply introspect. If an individual sits down, closes his eyes, and meditates on his mental problems, there is no doubt that that individual can cure themselves. This might work for small issues, but what about life-changing problems such as Alzheimer's? Disorders suchAlzeimers'smers and

Schizophrenia are inevitable processes of our bodies. Instead of trying to find a way out of all these disorders, one has to live through them. This might seem painful and unfair but things happened because they were meant to happen. Nature follows no doctrine, but humans, unfortunately, create doctrines and laws to enhance their lives. In reality, no human can be satisfied with even a perfect life. Satisfaction lies in the understanding of being calm in whatever nature provides. If one is blind, then that individual ought to live their lives blind. If one has schizophrenia then they must live through it. Surely their lives could be better if we found ways to enhance their perception of pain and pleasure, but in the end, it is only a perception.

The medical and the therapeutic fields help solve problems revolving around the five senses. We strive to make sure that we enhance our lives by using these five senses to the maximum amount that we can. But humans have to understand that even though their senses may be impaired, they have other tools in them. Just because they cannot experience the realm of perception does not mean their life is meaningless. Although individuals would rather live in the realm of perception, it is more healthy for them to live out of it. By living out the realm of perception, an individual isn't trapped by their senses. They are able to experience life outside the deception that is produced by senses. Senses belong to you, you don't need them!

Conflict of Science and Religion

For a long time, the majority of the human population believed in a divine being who dictated their world. Everything in the world was described as the god's creation, and there was no questioning about why things happen. The answer to most questions was: Because God made it that way. Also, technology and equipment weren't as advanced back then as they are now. So that brings up the question: Was science invented or discovered? Science isn't needed to survive, because people did survive without science in the history of humanity. Back then, people believed religion and god were concepts that were necessary for survival. People believed that their prayers would give them luck and that their gods would keep them safe if the people sacrificed animals for the gods. This mindset seems illogical in today's age, as most scientists only believe in that which is consistent and predictive. How do science and religion work with one another? Is it even possible for these two to describe reality? Or are they the same thing?

Religious institutions rely on other methods that are quite similar to inductive and deductive reasoning. There is often knowledge passed down through a book, which is written by a prophet who has been communicated by god. This is similar to deduction because both methods rely on some type of doctrine, or law that then narrows down the examination of truth. The notion is that no two religions are the same because their laws are different and these laws predispose individuals to know certain truth. Regardless of which religion an individual follows, the individual puts forth a supernatural being before himself. In science, the observer believes that he or she is capable of observation and coming to the conclusion of the truth. The basic difference in the two is that religion believes in an immaterial force dictating upon us through a supernatural being, while science believes that humans govern themselves and that in the realm of perception, humans are allowed to examine and come to truth by themselves without any other force needed.

Why was it then that no one practiced science in the early stages of civilization? Was religion something that occurred naturally without a name or identity? Is religion just a way of life manifested in different forms and names? The answers to these questions seem irrelevant to the scientific community because these answers are too ambiguous and vague. Nonetheless, the scientific realm can only go so far in describing how events occur through induction. Science cannot answer why certain things occur. For example, science can explain how sleep and arousal work but cannot describe why they happen. Physics can prove how changing a magnetic field produces a current, but cannot explain why it happens. On the other hand, religion can explain everything humans can think of. And the answer is simple: because God made it that way. This can allow religious institutions to blind people, but it serves to be a viable reason for answering questions that have no one specific answer. Science cannot deny the existence of god because it can't examine a phenomenon that lies outside the realm of perception. Nor can scientists approve of gods because they can't use reasoning or experimentation to conclude god exists. In the end, all scientists can do is ignore religious skepticism and continue to follow their methods of induction and deduction provided by their scientific founders.

Religion started as a way of life when humankind began. But as technologies and materialism advanced, humans started to believe that their existence is independent of any other non physical force. Science and religion are different and same at the same time. Science and religion both were produced by people in order to find a meaning to life. Both methods predispose the concept that there is an answer or that there is some truth to be achieved. But before seeking out for the truth, humankind has to question whether even there is some kind of truth. Or do humans just produce ways of psychologically reducing their cognitive dissonance and believing in something that happens to be consensive and consistent for the community? Both ways are necessary. Religion grounds the individual and allows the individual to become humble. While science allows the individual to be skeptical and build upon failures of others. Both systems allow humans to find something they are looking for. But both systems are looking for the same thing: purpose of existence of humans and space around them.

Is science needed?

Absolutely! Science is the foundation of our society and civilization. Although science may be biased at times, or uncertain and subjective at other times, it allows us to fly to space, transport to locations, communicate with one another and produce an artificially simulated realm(internet). One cannot sit down and worship god to fly to the moon. One cannot bend a spoon just by looking at it. Similarly, one cannot consciously live without using science. Science is quite paradoxical in the sense that it doesn't always seek an answer. The folds and the bends in the experiment serve as learning opportunities for further knowledge. And if science was just a mystical way of describing reality, then we, as a society, wouldn't have gone so deep into examining it. Science is a mere word to describe a process of understanding the world. But it is also a method of learning and producing objects that allow further knowledge of this universe. Regardless of how great science and technologies are, there are still mere parts of our perception. Nonetheless, they serve as very interesting tools to examine the physical world. Although we believe science helps us become who we are, it is important to not discard the fact that science only examines the conscious side of any individual. The unconscious side of the non physical side of humans and the world cannot be examined through science and reasoning.

    The process of examining something changes the result of that exact same thing. Experimentations and reasoning change the thing that the methods are trying to examine. Like mentioned in the knowledge chapter, the process of gaining knowledge and the realization of knowledge have to occur simultaneously. Because by examining the natural world through experimentation, we add nuances to the already complicated system. The process of science itself predisposes certain knowledge because it relies on the fact that everything has a cause and effect. The problem of science is that it can never truly understand a natural phenomenon by using human made tools. Those tools themselves deteriorate the process of gaining knowledge, thus producing a false conclusion. The only way of knowing whatever there is to know about the natural world is by being natural and introspecting within. Humans themselves are creations of nature. All humans have five essential elements within them. These are fire, earth, water, air and ether. The body, no matter how complicated scientists try to make it, is nothing but the composition of these five elements. I cannot propose this fact logically, because then that would mean I am no different than scientists. This phenomenon of element composition was discovered way before humans could ever read or write, or even think for themselves. How do I know this, you may ask? All through introspection and experience. Once you experience the process of being the observer and the thing that is being observed, you realize that the natural world is no different than your own entity. We are the exact nature that exists outside us. All we have to do is introspect and realize our natural place.

6. Concept of Arts

Subjective Reality

Art is a technique of expression and communication. Many concepts in our world do not have words, so art allows thoughts to be conveyed through sensory perceptions. Literature is one piece of art even though it uses words. In fact, the real definition of art is very confusing, art can be seen as perception of reality. The difference between reality examined by the natural sciences and the arts is that the arts emphasize the point that each one of us experiences reality differently. Thus each individual's art is a representation of their view on life. On the other hand, the sciences have a focus on a consensive rule about reality. Reality isn't subjective in the realm of science because there are certain physical and natural laws that keep knowledge precise and consensive. Education in science means that the teacher and the student learn content that is universal and is accepted by everyone. While education in art focuses on individuality and the concept of bringing thoughts onto visual or even other sensory mediums through different artistic choices. These choices are focused on guiding the artist and the viewer to better understand a certain concept. Since words are so direct to the point, they lack individuality. Thus drawings and visualitions are more helpful when it comes to conveying a message that is directed differently towards each individual. Thus art is not about the content, but rather the expression the content creates.

Literature

The word literature refers to the art of writing. This exact text is a form of literature as well. Literature is quite different from storybooks, or news articles. Although literature uses language and words, it never directs the reader to the exact message of the work.

Concept of Arts

Artists of literature manipulate their works to guide their readers to a certain message. This message is almost encrypted in the works of art, and the reader has to decipher the author's true intent. Many times, authors write on abstract thoughts that have no specific intent. The intent, all along, is for the readers to make their own interpretation of what the intent is. Some of the common ways of expressing literature are stories, poems, dialogues, theatre plays and even movies. The most basic form of literature is a story. A story, fiction or nonfiction, has a beginning, middle and end. But while reading, the reader is almost sucked up in the plot and experiences the story as if it were happening in real life. Because literature and the arts use imagination, they work in a realm separate from perception. It is true that all arts use sensory devices to portray the message. Yet that expression could revolve around guiding an imagination of some sort. Thus stories allow us to imagine and wonder. They allow us to question, for the first time, about reality. Stories are passed down not exactly as they are told, but rather how the listener interprets them. Thus literature evolves based on imagination and interpretation of the individuals who practice the art of expression. The word expression is being used a lot. Quite frankly, I can't use any other word besides expression to express exactly what I want to express. Expression is just a way of communication. Literature doesn't teach or tell something, it expresses a thought that has some literary devices to guide the reader to think in a certain manner.

Poems are another way of writing. Many say poems came before stories, while some say poems are stories themselves. Regardless of the history of poems, they are considered as very influential in our society. Just like stories invoke imagination, poems invoke emotions. The way poems invoke these emotions is through poetic devices. Devices such as tone, syntax, and rythme help the poet express his thoughts in a more precise manner. A poem carries a lot of emotion only if it has the proper literary devices to enhance the reader's ability to interpret the poet's thoughts. A poem without any poetic devices cannot express the poet's true thoughts because each reader will interpret the poem as per their mood and knowledge. To ground all readers to the same platform, poets use rhythm and tone and other devices. The devices serve as a key guide to the poet's art's true intentions. Many authors go above and beyond the realm of perception into the realm of emotion and imagination by using intense syntax and tone. The realm of emotion is different from the realm of perception because emotions happen without any much input. Emotions are quick reactions to a given stimulus. Authors and poets use the human's emotional side to bring upon many important messages. In fact, most religions have their sacred books in some sort of a poem or song. This allows people to bring emotion into their practice, which is vital to worshipping with a pure heart.

But emotions can also be exploited. Many songs in the 21st century revolve around depression, drugs and sex. These songs invoke a negative emotion of helplessness in today's society. Because people can use expression to invoke such a strong emotion in people, some choose to influence others in a negative manner. In fact the famous Shakeshphere wrote many tragedies and revolutionized the concept of literature. Much credit goes to Shakespeare for writing such tragedies to invoke a deep message about the concept of good and evil. The tragedies convey two messages: 1. Bad things are prone to happen to all beings, and 2. The nature of tragedies and fate is that people can't do anything about their life. These messages seem negative, and they are supposed to be negative to invoke an exaggerated concept of life. Tragedies are sad, but they are part of their lives, and part of the natural cycle of life.

Another example of use of negative emotion is when Franz Kafka wrote many books on the helpless nature of humanity and the inevitability of suffering. Later on, his works were branched into a specific category of literature called 'kafkaesque.' Kafka wrote a lot of great things, and the reason why they are great is because he used strong emotion in his works. His famous book called Metamorphosis draws upon the concept of irregular and unfair treatment of helpless people in the society. Yet his works are very similar to Nietzche and

Jean Paul Satre, both of whom also wrote about the negative aspect of our lives. There is no judgement of whether humans need negativity. The point is that authors and poets use quite a lot of negative emotions to bring upon a rebellious attitude. However, some expressions could be interpreted in an extreme manner in which some individuals could be trapped in the never ending negative cycle. Literature has the power to enlighten and express knowledge to many, but it also has the power to bring upon depression and helplessness in individuals. These emotions can be elevated to the point where individuals change their whole identity based on the intensity of the expression of a certain art work. Yet this is part of literature. Humans have to balance between negativity and positivity. Going to extremes can lead to over influence of emotions.

Philosophy is also branched as literature. Philosophy is different from poems and stories and plays. Philosophy doesn't have a structure of any kind. Philosophy can be as deep as anyone wants it to be. In fact, this book will probably go down as a philosophical work of art. But philosophy is still restricted to words, imagination, and interpretation. In the beginning of the book, I explicitly said that my words in these chapters are only guides, the true knowledge does not rely on these mere sentences. Any great philosopher knows the basic concept of limitation of communication and language. Before even talking about a concept in philosophy, a philosopher talks about the ways philosophy is taught and learned. Almost all professors and authors of philosophy emphasize on the history of philosophy. Because the whole point of philosophy is to break boundaries of knowledge, there are some philosophers who neglect the history of philosophy and previous philosophers. But, in my opinion, it is important to recognize the works of great philosophers(from east and west) in order to start the process of making and learning your own philosophy. The fact that these people gave their entire lives to philosophy means that there must be an immense amount of thought put into their works. But once again philosophy relies on interpretation so no matter how great of a book Plato might have written, if it cannot be interpreted correctly, then that philosophy becomes stagnant and irrelevant. In fact, many philosophies have been changed, manipulated, and even translated and in this process they have lost their true intent. This is a common problem of all works of art. Art revolves on the fact that it can be preserved and kept exactly as the author meant it to be. But unfortunately, things get old and any material or a piece of writing becomes either physically impaired, or becomes too distinct for contemporary viewers to interpret and connect with. Philosophy, in the context of art and literature, is ever so changing with time and people. Nonetheless, philosophy is a strong tool to introspect and start wondering about the nature of ourselves and our minds and the world around us. Philosophers can do the best to invoke this within their readers, but it is upon the readers to understand that the philosophy is only a vehicle of innovation, perhaps even a vehicle of transitioning from normal thinking to critical and imaginative thinking. All artworks do this, but philosophy is the best at it because it doesn't use emotions or any other senses. It simply uses the thinker's mind.

Music

Art is very hard to describe in words, but I am trying my best to use my vocabulary to describe art. Music is just one of those works of art that can elevate us to different levels. Music that is produced by any material at a certain frequency can change our brains and our perceptions. It is crazy to even think about how music is so engrossed in our culture, yet no one really understands why music is so important or what music really is. I won't get into how music is composed because music composition is way too complicated. There are over thousands of instruments and each of them has its own variations and so forth. But simply put, the art of music is heavily dependent on its variations and the sound it produces. The scientific explanation of music takes on the approach of simple harmonic motion. The fact that strings vibrate and have wavelength and frequency means that sound is variant on frequency and material. But instead of tangling into the realm of science, which I have extensively written about in the previous chapter, I want to explore music through a different approach.

Natural sounds of animals, birds, trees, water, and other things in the world happen without any interference from humans. So music isn't even invented. Music is one of the only arts that can be composed by nonliving entities. Accidental sounds have a rhyme to them as well. Music is not to be talked about or to be learned. Music is there, at all times. All we have to do is be attentive to it. Silence is a sound as well. Paradoxically, the sound of your breath is happening at a frequency that corresponds to the frequency of the environment you are in. That is why we are so tired after a concert because we have breathed extensively, and have lost a lot of energy. On the other hand, when we sit down in a garden and listen to birds chirping, the breeze wheezing, the leaves falling, and the river flowing, we naturally become calm and tranquil. One of the easiest ways of getting out of depression is by playing an instrument or listening to your favorite song. This is because music naturally propagates breathing which then either excites or demotivates the individual. Many spiritual people have the musical instruments that help elevate themselves to a different emotional level

Music doesn't have to revolve around nature, though. Our ancestors encountered music accidentally when they started hitting sticks on rocks. String instruments came out later, but humans were entertaining themselves with music long before any modern instrument was invented. Music doesn't need a medium, all it needs is someone who can listen to it. Just like our physical world cannot be perceived without eyes, the essence of music and its rhythm cannot be perceived without ears. The senses, no matter how flawed, can be used to do amazing things in this world. Music has the ability to change our thoughts and emotions. But music, just like any art, is only a method or way of expression. The expression is more important than the medium of expression. That is why no matter what instrument you use, how well you play, or where you are, if your mind is not interpreting the expression correctly, you are not understanding the message of the expression.

Modern Philosophy of Arts

The realm of arts goes beyond just writing and literature. Sensory arts are more common than literature because they take less time to produce and less time to understand. At the same time, they are also more subjective. The most common sensory arts are visual arts. Since vision is the most dominant sense in us, we try to experience as much as we can through our eyes. In fact, eyes can sometimes overrule other senses such as touch or smell. Most of the common visual arts have similar goals as literature. They mean to invoke a sense of emotion and bring upon a certain message. In fact many philosophers admired art for the fact that it brought up so much emotion. Sometimes art can bring more emotions than real people can, which is quite scary to think about.

Philosophers have pondered on why people chose aesthetics over normal or mundane objects. Can we even have a choice to like something that doesn't invoke emotions. We have already concluded that art revolves around the transmission of emotions. Thus emotions are the reason why we appeal towards aesthetics. Over a long time, we train our brains to feel good when we see a color or patterns of colors. At the same time, we can also condition our brains to feel gross or irritated. Regardless of what emotion we use to react, we inevitably associate aesthetics as something that brings positive emotions. Is art just something that is aesthetic? What is art? Many philosophers say human made art is not art, but rather it is only a representation of art. While contemporary artists say that art is anything that portrays a unique side of the artist. And then there is the famous phrase of 'beauty of any art lies in the eye of the beholder.' So is art natural or subjective to human perception and mind? Does art have any distinction from its artist? Does art have a non physical component to it? These questions are all ways of deviating from the concept of art, which is just expression. My theory of art is that art is a simple way of expression. The expression depends on the artist's artistic choices as well as the viewer's own interpretation of that expression. But instead of questioning what art is, or why aesthetics appeal us more, or why art evokes emotions, the more significant question is whether or not art is necessary or not. Can humans live without art? Is art the transcendent way to express ideas and knowledge that cannot be expressed by words? We arrive, once again, at the problem of language. And that is where I will stop this section, and explain the concept of art by expressing art itself.

Art to describe art

7. Politics and Economics

Breakdown of politics

As all humans have an intellect, they also have the ability to judge between good and bad. Yet regardless of this fact, there must be a system that governs human nature and guides them to live life according to certain philosophy of thought. Politics is associated with a system of human intelligence that governs the human community. The premise of all political philosophies relies on the concept of self interest. Regardless of whether self interest is good or bad, it is a vital part to understanding the foundation of politics. Politics can be broken down into three things: power, morals, and doctrine.

Power is the most important foundation of any political structure. There are two components to power: who has the power, and who is the power being used on. There have been many philosophies on what there needs to be someone in power, and most of them argue that power keeps a hierarchy in which the concept of being on the high platform is seen as a motivation for all in of those in the system. If there is power given to one entity or one group, then that means the system is a totalitarian state. Most of the monarchies in the initial state of civilization were monarchies, who had kings and queens, and the people of the kingdom were to follow those in power. Yet as civilizations improved, they realized that a political structure shouldn't just revolve around one entity, but rather the entire civilization. We can talk about how unfair monarchy and totalitarianism is, but that is because we live in the 21st century and because we have the ability to question and wonder about justice. Back when monarchy was present, nobody thought of any other way of society. Soon power was given to sections of the society, and these sectors were to enforce law upon their people. As time passed, power became the topic of debate and more and more people started questioning why only some should have power and others shouldn't. Socrates and Aristotle have their books written on their stance on the political systems. Each of them says that power must be in the hand of the intellectual one. Monarchies and oligarchies still existed in the world, but democracy was more predominant than any other system. Yet just as democracy started to expand, opposing systems came into existence, and soon there were three wars based on ideological and political divisions. The Cold War was the last among those three where democracy won, and as of today democracy leads as the most balanced and efficient way of governing the human kind. Democracy gives the power in people's hands. But there was meant to be a social contract that came along with this. The concept of social contract comes from an English philosopher, John Locke. Democracy was exciting because everyone had power, and everyone had a voice. The history of power goes from one to many. But all power is judged by morals. So how do morals play a factor in the government structure.

Power and morals go hand in hand as power controls morals and morals control power. Humans have always believed in good and bad, The justification of good and bad is something that we have built within us, thus we already know how to judge others based on a moral belief. In the beginning of civilization, all morals were based on a sense of divine being, or God. Religion and sacred texts became the foundation of morality and soon people believed that good and bad were associated with their intent rather than actions themselves. This was quite different to the ideological or utilitarian way which proposed the concept of creating the maximum amount of happiness, regardless of intent or the harm in producing that happiness, so long the benefits outweigh the costs. At last, free will was questioned, and humans started questioning themselves and wondering if they are good or bad. Thus, political structures no longer only cared for maximum efficiency and outcome, but also worried about human welfare and the act of morality. Long ago, kings were symbols of God, and in such a manner, their say was the only way to live. But this fell apart as humans started questioning whether God truly existed or not. In this manner we went from a system that produced morality strictly through the means of what one entity said, to a system that allows everyone to choose and decide morality and come up with their own justification for their beliefs. Yet the government was highly involved, even when people were given the ability to think and freely choose their laws of morality. No system is flawless, for there are always individuals who bring upon revolution to break the system, and find ways to convince people of a better system. Revolution is inevitable so long humans try to find self interest and a desire to bring change. Nevertheless, morality serves to highlight as a keypoint in any given government. If there are no morals, or no ways of deciphering good and bad, then the society fails to pursue anything or have a purpose in life. Thus morality, whether it is imposed or brought upon from within, is necessary for humans and will always be there to guide humans.

At last, there is the doctrine. The doctrine is the physical form of the ideological concepts brought upon by power and morality. Not all civilizations used a doctrine, but there were ways other ways of passing down the political system to upcoming generations. Each generation thought that their political system was the best one, and so they tried to spread that message to their upcoming generation. Obviously, the political structure changed over time, but the political ideology was stored and improved. A doctrine of governance portrays the necessity of keeping things organized and allowing structure and regularity in the society. If there were to be chaos or any sort of crime, then the doctrine would provide the rightful answers to implementing justice. This way of solving justice pre assumes that humans aren't capable of making rightful decisions at a moment due to their emotions and impulsions. But isn't it so that humans were the ones who decided what is good and what is bad? Although, there might be an argument about god's will and destiny, in this case, it was the humans who came up with the doctrine. So the question then becomes why is it that those who came before us decide what is moral and what is immoral. Although democracy allows individuals to act according to their free will, democracy doesn't permit anything that lies outside of the constitution. Law and order are necessary to welfare and to implement this, we need a doctrine which has to be in the process of constant change according to time and location.

It is through these three pillars, that any government is able to continue its governance. If at any time, one of these three is lost, then the act of governance is also lost and the system becomes fallible. No matter how truthful or loyal a government is, if all the power lies in the hands of the government, then the act of governance will soon stop due to revolution or some sort of change. No matter how much power is in the hands of the people, if there are no morals to live by, people will not know their purpose and soon give up everything they control. Yet, the use of doctrine is a little different. People can live without having a doctrine, but the doctrine makes law and order more impactful and compulsory.

Altruistic Social Anarchy

Now that I have considered the history of politics and governance, I want to propose my ideal way of life. Since this topic is so debatable, I want you to know that this is only my opinion on what is best. Clearly, you do not need to agree with me, but please be open to my idea. My approach to politics lies on some presumptions. I believe that altruism is possible and that given the right environment people could behave altruistically and freely simultaneously. Secondly, I believe that anarchism is the ultimate way of freedom in life. My version of anarchy is a little different than that of the modern society. I also believe that education is the key to getting from democracy to altruistic anarchy, and finally the process of moving from democracy or any capitalistic political system to anarchy lies in the concept of natural process. Like Karl Marx, I also believe that there mustn't be a forced revolution to create this political system as people will have the competency to do it by themselves.

Let's start by talking about altruism. What is altruism? Altruism is the behavior of doing something for someone else without expecting anything in return. Or in other words, it is the act of being selfless at one's own free will. In the beginning this act of being selfless may not seem like free will, but that is only because we have a predisposed concept that humans are selfish and we need to protect and care about only that which is ours. Before any type of government structure was placed, humans lived without worrying too much about living. Darwin's theory of survival of the fittest is only a theory, may it be good, but is only a theory of explaining why humans evolved in a certain way. Darwin forgot one important concept in evolution, which is love and intrinsic happiness. When I say love, I do not mean romantic love or any sort of sexual or biological theory of love. Love is a natural process of caring and being compassionate. It seems illogical to believe that someone would give up their survival for the sake of love, but that is how far our intellect has gotten us. At one time in history, people didn't worry about good and bad, as they were intrinsically aware of good and there was no way they would do anything bad. Altruism was not just something people did, but was inbuilt within them. It defies the concept of logic, but so do half of the theories about galaxies and black holes. The reason why altruism works is because it revolves around the concept of welfare of everyone. Democracy believes that it brings welfare to everyone, but it truly doesn't as there is still sorrow and depression all around the world. The crime and suicidal rates have only been going up in the 21st century. Democracy is improving efficiency and technological advancement, but it is not improving human life. Democracy is great, but there is a step above it. That step cannot be reached through capitalism. Hence I propose anarchism.

Altruistic anarchy relies on the fact that everyone has free will of helping others without a sense of wanting returns. Although it may seem too idealistic, the concept of anarchy allows individuals to do daily activities on their own free will. There is no political system, as there is no power structure. There is no doctrine or moral code that defines people to live in a certain way. There is no law or order to make sure people have a purpose. All these might seem extreme and necessarily idealistic. Yet it is important to give people all their freedom. The human will have the chance to create their own morals, and believe in their own religion, or even make one if they feel like it. Each human will be unique, but that is what humans were meant to be. Freedom will allow great amounts of thinking and intellect to advance. At the same time, freedom won't be exploited as well because humans will help each other at their own free will. Humans are capable of such great systems because their natural form is based on love. Altruistic anarchy will lower focus on technological and materialistic advancement, and focus on human development and spiritual exploration. There need not be any way of forcing people to do anything, as they will follow this system as if their lives depended upon them.

The altruistic anarchy system would ensure people that so long they helped others, others would help them inevitably. Though this concept has some selfishness, it is required for human motivation and growth. Humans will not only gain intrinsic happiness but will also inevitably gain extrinsic happiness. Humans won't betray the exact system that grants them everything they needed and wanted in their lives. This system relies heavily on the fact that humans will help one another, but that concept will be natural due to the environment provided. It will be sufficient for a maximum amount of people to be altruistic, if not everyone, as those who are not altruistic will fall behind and would eventually want to blend in with the society. Humans are social creatures, and psychologically we are prone to fit in and make the best of the environment provided to us, whether that be an environment that pleases us or that which doesn't.

No system is flawless, for there are always individuals who bring upon revolution to break the system, and find ways to convince people of a better system. As long as the majority of the people act altruistically, the minority will feel uncomfortable or even unsuccessful in bringing upon a revolution. Yet there will be people who continue to follow the system and continue to support these individuals and try to help the minorities. The system is based around free will, and if minorities exploit that system, then they will also realize that they are putting their free will at risk. Thus no one would exploit the system that gave them that exact opportunity of exploitation due to free will. As a result, most people will realize that true happiness lies in giving happiness to others. In this manner, everyone in the system will be happy, and no one will feel the urge to revolt. Yet there are abnormalities in brain function, and thus to fix the problem of psychopathy and criminology, proper education will be provided.

Education will be the foundation of this system, as proper education will be the way to switch from capitalism to anarchism. Humans, if conditioned properly and educated properly from a young age, will act as per that condition only. Education won't be assertive as in brainwashing the people, but rather enlightening and eye opening because it would only invoke a natural response of altruistic behavior within the children. As Locke stated, there must be a social contract in order for a society to function, I also believe that if people want to enjoy their freedom and live in an altruistic system, they must give in to their education. The first eighteen years of everyone's life will be in the hands of specialized individuals who show the highest amounts of selflessness in community. How these people are selected remains a topic for another time, but the criteria would be based on their altruism and a proper study of the art of education. Although this seems like a controlled way of teaching, the students will exercise the greatest amount of free will. There will be one condition, however. And that would be that the students will stay at school, seperate from their parents. Although the students would be free to go around the world, it would only for a while. As it seems, the education system will be strict, but it will be a payoff for the upcoming adult life. This is the only way anarchism in the context of altruism would work. So long as people are not foolishly following their emotions and corruptive impulses but rather are guided by intellectuals that are highly altruistic, the system will continue to advance and be effective in reducing revolutions. In its final form, the system will have no teachers, as all parents will be their guide to their students. Yet this is only possible when revolutions decrease in number, and everyone in the society has adopted the new way of life. The curriculum of schools will be the only thing that the students will not have freedom upon. Because in order to have the freedom, the students must first learn how not to abuse it through education.

In these ways, the political structure of altruistic anarchy will create maximum amounts of happiness. It won't even be a system, as it will be a normal way of life soon. But in order to achieve that, there must be some changes in the economical status of the society.

Economical de-construction

The word economics is associated with production, distribution and exchange of goods and services. There is a whole concentration of industry that focuses on marketing and supply management. It is through capitalism that all economical facilities are able to thrive in the world of democracy. Capitalism is the concept in which trade and industry companies accumulate wealth and keep it to themselves without the intervention of any higher power such as state or nation. Inevitably, capitalism promotes a free market which invokes drive to materialistic gains. In an altruistic system, there won't be any sort of monetary exchange, as people will gain and provide whatever they want as per their need. Although there would be companies and job opportunities, they would all be based on providing aid to the needy ones. As technological advancement will slow down, the materialistic desires will also lower, which would mean that people wouldn't worry about money any more. The focus of life would be to a. Teach children how to be altruistic, b. Help those with mental and physical issues, c. Find ways of survival means, and d. Explore spirituality. These are only guides, as it would defeat the purpose of anarchy if one were to tell others exactly what to do. Accumulation of wealth wouldn't be a concern, as exploring natural life and survival would be of greater issue. The self interest and ego would soon fade away, as individuals would live in communities, and help others all while living their own lives the way they want. Production and exchange of goods will happen at a mutual level, as each individual will determine the value of whatever the product may be. It won't be much of a barter system, but close to it, as people won't trade things, but provide things at other people's necessity. Exploitation wouldn't be a problem as altruistic people will cover up for rebels in the society. But like all systems, this system will have punishment methods too.

Instead of any imposition of a punishment, the individual's guilt and shame will be the punishment and a reminder for that individual of his crime. The shame and guilt will come from within, as no one would say or do anything to them. When an individual doesn't feel shameful or doesn't change their behavior, which defies the concept of being selfless, then that individual will be punished by his own community. The punishment will be decided upon their own decisive methods. Though punishment won't really be expected as individuals will be highly educated and will have a proper sense of justice. Yet in cases where punishment does occur, that criminal would have to live in isolation. It would be quite similar to prison, but they will have time to reflect and act and think on their own. They will go through another educational program which will once again try to lower their chances of rebellious action, and invoke a sense of wanting to do good for others. Once again, the only concept that will strip away free will is the education, which is geared to providing that exact free will in a rightful manner. In this way mutual trust will be accomplished and more people will forget about revolting and just continue to live naturally. Even though this system is designed to be a utopia, it will result in less of such results. Yet, it will still show the power of altruism and true nature of humans. When given free will and educated to not exploit that free will, most humans will learn to love and care and be happy.

8. Concept of Emotion

Emotion and Feelings

Emotions are the building blocks of our memory systems, and act as one of the most important tools for humans to connect and interact with one another. If anything, emotions play a vital role in our psychological lives, as it penetrates through intellect and knowledge and enforces its own influence on the individual. We talked about emotions extensively in the chapter about arts. But that platform was too narrow to examine the influence of emotions. Because emotions are inevitable, they are quite hard to examine, because the act of examination itself can be stimulated by emotions. It is important to distinguish between emotions and feelings, as they are completely different concepts. Feelings are generated by instinctual thoughts, while emotions are generated by reactions to external information. As long as we allow feelings to arise as a product of our own intellect and instincts, and not that of others, we can safely state that we have rejected mood.

All emotions arise in a very abrupt manner. Most emotions can be watched over, unlike feelings. Feelings are results of our own judgement, as they try to find some way through our good and bad thoughts. On the other hand, emotions are by nature results of some reaction. The word emotion comes from the word emote. To emote is so respond or act in a particular mood. By nature, our reactions are measured through the way we emote. Thus emotions are more reactive, and feelings are more interactive. Though the two seem different, they are quite similar in the fact that they both produce a sense of happiness or security for people.

Natural vs Unnatural emotions

Emotions can either be natural or unnatural based on their method and context of usage. Some examples that arise from unnatural emotions include self imposed techniques such as drugs or drinks. Unnatural emotions are brought upon to gain some sort of pleasure or get rid of pain. Pain and pleasure are complex emotions that cannot be reduced to happiness and sadness. Some might even say they are results of chemical influx or hormone reaction in our brains. Nonetheless, we can break them down as inhibitory and exhibitory reactions.

All drugs are classified as either inhibitory or exhibitory. They either reduce or increase the intensity of neural signals. Yet, we can become tolerant to certain chemicals, and soon our brains might think the inhibition or exhibition is noramility. Similarly, any emotion, after a certain time period, seems normal. Yet these are all signs of unnatural emotions. If emotions are received on the platform of any substance or outside chemical, or if they are purposefully produced to increase efficiency in us, then they are not natural. Furthermore, if a person regulates their emotions for their own desire then they are decreasing their chances of getting that exact emotion.

Natural emotions are evoked due to natural tendencies of our sympathetic and parasympathetic nervous systems. I have not commented on whether emotions are good or bad, which is subjective itself. The question isn't whether we should have emotions or not. It is quite impossible for any human to leave his or her emotions outside their daily lives. Yet we should watch over our emotions, and be cautious of them. If our tendency is to disable our intellect and flow with the temporary mood evoked by emotions, then we cannot find truth or happiness in anything. Yet if we are more likely to know our emotions, and know our tendencies, then we are able to act accordingly when the emotion finally comes to us. Natural emotions do not occur according to expectations though, so one must be prepared for any and all types of emotional stimulus. As emotions are only reactions, we have to react with awareness, so when the same stimulus occurs once again, our brains are more capable of finding it unexpected and thus they are less affected by them. This same mechanism is present in our immune system, where plasma b cells that are produced in the bone marrow, specialize into memory b cells to remember the antigens on the pathogen. Our bodies are good at remembering obstructive and unnecessary information so that they can protect us from mishappenings next time. Similarly, our emotions can become less effective on us if we are aware of them. In this way, we won't allow our emotions to control us.

Handling Emotions

Emotions can be detrimental to our well being as they change our perspective and understanding of ourselves and others around us. Thus, we must make sure to handle emotions in a careful manner. We mustn't be emotionless, for that would be the path of nihilism. Instead, we have emotions, but not let them drag us into unnecessary realms. It isn't my prejudice that I claim emotions are harmful or unnecessary. For emotions can bring great amounts of joy and peace to many people. Yet, my point is that emotions mustn't be allowed to act without attention. When we are happy, we must remember the cause of our happiness and also remember the impermanence of it. We shouldn't feel depressed or anxious about the fact that all emotions are impermanent, rather we should be happy to be able to experience everything the world has to offer. The controlling of emotions comes in four parts: 1. Reflect on a previous emotion, 2. Try to recreate or be in a similar emotional evoking situation, 3. Remember your previous reflections and try to be aware of emotional arousal, and 4. Instead of suppressing those emotions, redirect your attention to your reaction.

Try to not react in an inappropriate or unnecessary manner, which you would otherwise act upon due to emotional influence. This simple theory of reconditioning the brain to face emotions as natural phenomenons and not not reacting to certain arousals will keep anyone calm. Soon, the practice will be focused on being equanimous, which is a state of being in which an individual loses any attachment to certain arousal of emotion, and acts equally and calmly in all situations. In this manner, the individual becomes equanimous, where all emotions, no matter how different, are handled in the same manner.

Some people say 'fake it till you make it' and although this idiom is quite popular, it can be false at some times. We can try to fake our emotions, but that is not a good way to handle them. Most therapists will recommend to smile more because it strengthens the neural pathway between your peripheral nervous system and the facial muscles. Physiologically, it makes sense to fake laughter or fake any emotion for that matter until you have mastered the art of deceiving yourself. Yet this could have side effects as losing individuality or perhaps feeling even more worse as you start to realize your own deception. The proper way to handle any type of bipolar disorder or depression is not by faking your system, but rather by observing it and allowing emotions to flow through. Although catharsis isn't the best option, having a dairy or a journal to write out all the miseries and pain could really help. At some threshold the patient will realize himself that their emotions are getting out of control. It is at this point, that the patient must start the practice of introspection through meditation. Mindfulness has already shown to be one of the best methods in therapy. And people are starting to look at it not as a spiritual or a religious practice, but rather a healing and psychotherapeutic practice. Thus emotions must be reflected upon, observed on, and atlast conditioned to be viewed as natural occurences to all human beings.

There are a lot of things that are beyond human control. No matter how much technology advances, we will still not be able to control creation, sustenance and destruction of nature and of ourselves. Emotions are one of those concepts that cannot be examined much further, as they are abstract and are out of our control. Our impulses are naturally occuring, yet it is possible through practice to lower unnecessary reactions that would otherwise cause abruptions in our daily life.

Expectations and Desires

Our expectations and desires are often causations for our emotions. We expect life to occur in a certain manner, and thus we try to create a picture of that perfect life in our heads. When that expectation is met, our emotions arise and soon we cling to that emotion. Dopamine is the neurotransmitter that is released at the moment when an event is about to occur. We get excited before the event even begins, as we anticipate emotions to arise. It is never the event that pleases us, but rather the emotion of reaction we get out of it that gives us the pleasure or happiness. Thus expectations, if too high, can also lead to unnecessary disappointments. Just like expectations are just a prediction of the future, so are our emotions that take place before the action occurs. If things go wrong, and our anticipated emotions do not fit the event, we get disappointed and feel lack of motivation. Such events can be dramatically worsened with the use of drugs or alcohol that penetrate our intellect and further leave into the realm of emotional influence. Thus one mustn't expect anything with a strong passion, and even if one does, he or she shouldn't be attached or dedicated to being happy or sad due to that expectation coming to reality. All that happens in life is a product of an unknown, yet not disproven, natural process. Even if we detach from our expectations, we become less likely to react due to our natural emotions, ultimately reaching a point where we are equanimous.

Detachment from desires

Similar to expectations, desires also act to create emotions. Desires are far more subtle, yet powerful, than expectations, as they reside deep within us. Desires, themselves, are not detrimental, but it is the attachment to that desire that leads to turmoil in emotions within us. Emotions derive from such strong desires, may they be influenced naturally or unnaturally. By creating a mindset in which our desires are controlled, not suppressed but controlled in a manner in which they are not allowed to flow freely within us, we can become more satisfied and calm in our lives. Most desires come as a temporary means of happiness. An individual wants an object only until an upgrade is available. Such a system evokes temporary desires that influence our emotions. If only we could give up our temporary desires, we could receive a great amount of satisfaction for a long term. As desires are a product of human nature, we should not suppress them. Treat yourself once in a while, but do not wish to continue to treat yourself on a daily basis. Have desires of becoming great and wanting to do great things in life, but do not become attached to the feeling of being great, as greatness is only subjective and soon will turn into nothingness when there is something greater to achieve. Thus make sure to detach yourself from the desires, and allow yourself to observe them as natural occurrences, just like emotions. For as long as humans continue to feed their temporary desiring minds, they will never feel the satisfaction of long term happiness. The cycle of emotions and desires will forever succumb them as they will continue to run after unnecessary things and gain a temporary emotion in return.

9. Concept of Compassion

Evolution

Love and compassion are usually interchanged and used as the same concept. Although these two are similar, they are quite different in many ways. In this chapter, I will explain how love and emotion evolved over time. Also I will be explaining how compassion is non preferential love. As Kierkegaard said in his book *Works of Love*, the man loves someone else for the sake of loving himself. The argument goes like this: A man falls in love with a woman not because he wants to make her happy, but because he wants to make himself happy. By creating a relationship, the man indirectly suffices his own greed. In other words love for others is just a cover for self love. Although this argument seems extreme and pessimistic, it serves to tell us about human nature and our centralized ego systems. A man needs to love himself more than anyone in order to survive, but he also needs to reproduce and pass down his genes. How does this all happen? And what is the connection of evolution to love and compassion?

There are many benefits to loving one another. Humans are social creatures that need a group to survive. Humans have developed a special kind of method that helps them survive. That method is language. Language makes our survival crucial as we can express our intentions and signals very clearly. Without a group, we won't be able to do that, nor would we be able to survive for long enough. We are apt to hunt and gather together as that is one of the most efficient and safest ways of living. This is not only present in the human community but also in the animal kingdom. Most animals live in groups, making sure they benefit one another. The concept is that each species tries to help its own kin and in return gets a protection warranty. This is a genius as natural selection tends to favor those who are strong and capable of sustaining themselves and their offsprings. Since we live in groups, the chance of the offspring surviving even after its mother has passed away increases. By and large, evolution has favored those who live together, and inevitably humans found out that picking partners and staying with them the rest of their lives is the best way to protect and raise their children.

Primal acts of love were very different from the romantic scenes we see in today's movies or shows. Humans do show emotion and self control at times, but nature has made it so that our sexual instincts are very impulsive making sure that reproduction occurs. Hormones such as testosterone, progesterone, estrogen and oxytocin all evoke this sudden arousal in a human that promotes sexual reproduction. This, in my opinion is inevitable because if reproduction was not promoted through homonic activity, populations would have fallen, and the cycle of life and death would have become stagnant. In order to continue the cycle of life, to bring new organisms, to support natural selection and process of elimination, mother earth has created strong chemical activity that promotes such reproduction.

The story doesn't end here because it isn't just the hormones that play a role in allowing reproduction to happen. It is also competition and competency. The female organism chooses a male organism that is suitable for her offspring as she wants to make sure that her offspring has the best genes available. This competition is built in nature and balances life. Once again, natural selection shows us as only the males with a particular behavior or competent ability are able to reproduce or find females.

Our primal and animalistic instincts are very in built in us to make sure that the cycle of life and death is not disturbed. All organisms are dependent on one another, and even if one organism's reproduction and population levels increase, the entire system falls apart. But humans have the intellect to control their instincts. Nature has also given the human the ability to self control these sudden impulses. As they say, we are nothing but products of our environment- social and natural. We are natural beings, mended and taken care of by nature. The fact that we have such a complex brain and the fact that we are able to restrain ourselves perhaps against the natural phenomena of sexual reproduction only makes me wonder about its purpose? Why are humans non instinctive animals? Rather, why are they able to restrain themselves? If we really wanted, we could completely stop life on earth at this moment, but we don't. And that is because we always balance our rationality. We say we should self control not just sexually but all kinds of other tendencies. But we also say that we must not restrain them too much as that will only end life on earth.

Love and Compassion

Now that we understand that humans are both instinctive and self-controlling animals, let us look at emotion. In the previous chapter, I discussed emotion as transporters of our desires. But not all emotions need to be criticized. Although emotions can lead us into dismay and chaos, they can also lead us into joy and bliss. Love is one of these emotions. I would go as far as to say love is probably the root for all other emotions. When it comes to love, we don't associate it with sexual love or impulsive love, we associate it with gentle and caring love. This kind of love is first seen in the child and his mother

There is a very peculiar and warm kind of love in children and their mothers. Often because the mother is the first person to take care of the child. Also the mother does not look at her child as another being, but as a part of her own self. Studies have shown a rise of oxytocin density in the mother's hypothalamus when the child is happy and safe. On the other hand, these same levels go down and cause stress hormones to be released when the child is far away from the mother, or is crying. This pain is felt by the mother, and only by the mother. Nature has allowed such a strong bond to only exist between mothers and their children because it is crucial for evolution and child development. If the mother is uncaring and does not see her child as part of her, she is more likely to first feed herself, take care of herself and then look after her child. This process is too inefficient. Patricia Churchland, a neurophilosopher, has a very interesting book called

Touch a Nerve, in which she talks about how the mammalian brain is very immature at birth because it needs to learn and develop that brain due to its environment. Since the baby is so immature and unable to protect itself, it needs the mother by its side all the time. The only way to promote this is by making the mother feel pain and pleasure of the baby, thus making sure the mom takes good care of her offspring.

We can go on talking about such evolutionary patterns all day, but the gist of all this is that this true, genuine, warm hearted love only comes when the mother *feels* as if her child is part of her. As I said earlier, humans are very self centered organisms, yet they have potential to develop empathy and compassion towards others. It doesn't always have to be the hormones or chemical changes in the brain that create such compassionate behavior. That is only there to promote survival and development. However, survival isn't enough. Nature has only made it so that the cycle of life and death is intact. Yet it is our own duty to make sure to keep the cycle of compassion going. In other words, nature provides the basic necessity of existence but humans are the ones that have to enrich this existence by creating habits for empathy and compassion.

Primal and sexual love or even romantic love is not the same as compassion as it involves self love. The Kantian philosophy of contra causal theory suggests that you are truly able to self control yourself and do good for the sake of others. But clearly, modern neuroscience has disproved this and personally, I think Kant is off the mark here. We do have certain levels of self control, but once again, we are abided by nature's predispositions. So then is compassion even possible? Can we go even further and ask whether altruism is possible? I talked extensively on altruism in my politics and economics system. Many would critic my theory and say that humans can never reach that level of selflessness. Or perhaps they might argue that we are bound by our animal instincts and that we can never overcome our egotistical nature. I agree that pure altruism is never possible as they say, but each human is able to conceptualize or comprehend some sort of altruism to practice. In other words, the concept of altruism is important rather than its detailed attributes. And I believe given our ability to exercise self control and compassion, we are able to have some level of altruistic nature within us.

Just like a mother can feel such an intimate connection with her child, though it may be due to nature's predispositions, we learn how to do the same with everyone else. Selflessness is not cowardness and I am not proposing self-sacrifice for someone else, though some noble people in this world would go even to that extent. Nonetheless, I think we can habituate ourselves to connect with one not with others, and feel joy and pain accordingly. Once again, though, we mustn't allow those emotions to strangle us and keep us worried. We must act upon those emotions, to make sure that those who are suffering, or those who are under stress can recuperate from their illnesses. How do we do this? How do we shift our natural attention from ourselves to someone one? Love allows us to do this to a smaller extent as we work for our family and do things for our family. Even some nonprofit organizations allow us to do this as their job is fully focused on raising funds or providing meals for the needy. Even while we are on the bus, we make sure that the elderly ones get the seats, even if that means we have to stand up. If we are at a hotel, we make sure to keep hold of the door for people behind us. At librarie,s we keep quiet for everyone's sake. At a restaurant, we leave tips for the waiter, knowing that he is slightly underpaid. All these 'manners' as we call them or conventional behaviors of morality ground us and keep us under control. You wouldn't see a monkey opening the door for someone or having the conscience to let the elder ones sit down. But humans can understand one another, to control ourselves for the sake of others. Surely we aren't bounded by animal instincts, we are way beyond that

This all leads me to say that humans have the potential of habitualizing themselves to do good for others. All behaviors mentioned above are simply common sense for us, but 500 years ago some of the exact same principles would have been absurd. The point is we are, and we have been slowly progressing to become an altruistic civilization. In some regards, we have been doing excellent work at this progress, but I still think we could do better. Instead of having conventional wisdom, or 'manners' according to the situation, we must develop the concept of altruism in young ones. Children only imitate their surroundings, thus all of us have to be more selfless. Culture can drastically affect the way people behave in society. We see this as some of the african tribes, to this day, practice nomadic cultures and live in the forests. Their children, if they were in civilized nations, would not know how to act properly. And this is not because they are 'bad' or 'shameless' but because they are psychologically not developed in such an environment. This goes back to the education section. We must condition the youth to become selfless, to do good for others. In this manner, it will be common sense for all humanity to leave hostility and suspicion of others, and bring forth compassion and love.

Psychological benefits

Mental health and psychological treatment has been only severe lately. The disorder levels have been going up, people are becoming more depressed, in general high schoolers are facing suicidal moments, and children of all ages have become unbearingly selfish. Of course this is the broad view on the mental health issue, but if we don't pay attention to it or don't change something in our society, youth is only going to become more and more dissatisfied in their lives. One of the primary reasons for even non clinical depression is the fact that children are being raised in a society that values money over virtue. I have already mentioned in earlier chapters how the education system is geared towards making sure children get jobs, not geared towards allowing children to explore and learn more about themselves and others around them. The IB programme is a good option, yet even that has been stagnant over the last couple years. Why are more people facing personality disorders? Why are more people depressed even after getting everything they want? Why are high schoolers going through suicidal thoughts even though they have comforting friends? The heart of the problem lies in the fact that we have lost purpose.

Many teenegers have complained about having no meaning in life. Maybe that might be just me or my friends, but I feel that students are becoming more and more understanding of their situation. They all strive to achieve their goals in the future, but while doing so forget the most important part of life: compassion. Narcissism disorder has been increasing all, providing reason to why people are so selfish. It comes from the concept of consumerism. So long as consumerism is a main ideology of the society, mental health issues will only get worse. Accumulating wealth and securing a job position does not lead to happiness, nor does it lead to any sense of true meaning. It just makes you fall into the non ending cycle of going to work, getting money, and consuming products. To make things worse, the market system is only promoting such behavior as items are sold left and right just to attract youth to buying things. As if there weren't enough things to buy, companies are becoming more and more competitive and hostile towards one another and thus producing more entitling and attractive objects for people. This just creates a loop of never ending consumerism. On the other hand, there are poor kids who are striving for two meals a day, who have no education, whose family is addicted to drugs and has lost meaning in life. Such people exist as well and believe it or not those consumerists, with great jobs and salary, are prone to share the same situation as the poor ones one day. Money does not solve problems, it only makes new ones. Love and compassion solves problems. Selflessness solves problems and brings welfare for the community. Yes, there are many social organizations that try to help such needy communities. But the deep rooted problem is our capitalistic ideology. We can't rely on consumerism to make people happy. Though it helps our economy, and makes the rich richer, we have to have some empathy towards the poor. For this reason, altruistic anarchy becomes a great option. See previous chapter for detail.

Nature of Reality

Compassion brings people together, it removes the hostility in the environment. Psychologically, we are more healthy and wealthy when we not only receive love and care, but also give it back to the people around us. Such intrinsic gratification will only help everyone achieve what they want in their lives. Social psychology and positive psychology has proven that happiness is a product of giving and not taking. Suppose I told you that you would be more happy and satisfied by giving five dollars to someone else than by using those yourself. You would think no human will ever do that, and if they were to do it, they wouldn't feel any different. Wrong. Habits of compassion and willingness to give to others only makes oneself happy. After all, we all are connected with one another, why would we have a problem helping ourselves by helping others?

Practically, I know that though there is no such thing as a selfish gene, humans are apt to be self preservatory. Yet, I think proper conditioning, habitualization of self control and utter conviction of being selfless will ultimately raise our current society. It isn't just the economic or political system that has to change, it has to be people's willingness to change as well. But unfortunately, today's generation has its feet in the mud as they only care about themselves(generally speaking). It isn't their fault either as they were told that true and genuine happiness lies in ourselves, in our morals, and in our ability to help others. The Buddha taught so many great lessons about self control, compassion, pure will, and inner peace, yet the western culture has never seen them as practical. It all stems from one misleading ideology: consumerism. If we could accumulate all the land and wealth that there was, what would we do with it? It isn't the material or wealth that the man is after, rather it is the anticipation of getting more. To continue to find meaning in consuming unessential things simply for the sake of it.

Peace, a term used everywhere at all times. Is it possible though? Will it take us anywhere? Or will the powerful ones only continue to take advantage of the powerless ones? Unfortunately nature has made it hard on us by giving us a brain that looks for pleasures and expects respect. But nature has given us the potential of changing our root core beliefs of ourselves. We may think it is never possible- never possible to be compassionate due to our self preservation instincts, yet I propose that it is possible. All we need is a strong conviction. A conviction of being more aware of our consumerism. We are to suffer psychologically, if this continues. Divorce rates, crime rates, mental health rates have only been rising and they are all rising for one reason: inability to find love. If only one could give up their foolish ego, their unquenching desires that provide little to no true happiness, maybe one day we can have a society full of compassion. Full of genuine love- not the kind of love that is seen in romantic movies or in TV shows, but a love that allows us to look beyond us. To feel for others, and to protect and respect others. If there is such a thing as peace, it will only come in a subtle form- where the society is connected by conviction and morality.

10. Concept of Memory

What is memory

If I remember correctly, memory is the recollection of an object, person, linguistic phenomena, or combination of all. Oh, that is only correct if I remember the definition and concept of memory quite well. Similar to language, memory is an odd way of knowing. But unlike language or sense perception, memories are very abstract, unknown and vivid. As previously said, memories are very closely associated with our emotions. We remember things if they hold a strong influence on us. But that doesn't mean that all memories are products of conscious thought. There are some memories that are formed unconsciously, perhaps in our dreams. There are some memories that are formed consciously, but then altered unconsciously, and then recollected as complete different memories. But there is one concrete thing about memories and that is that they always come from our past. One could argue that someone with deja vu could remember things he never did in the past, but are things he is anticipating to do in the future. Nonetheless, these people are very few in numbers, and even they themselves have very few encounters where a dream of recollection of memories becomes their future. I am not talking about psychological hallucinations, I am talking about actual memories. Memories of this world- the physical world. Memories of the mental world are way too hard to understand. Nonetheless, I will touch on mental memories later in this chapter.

Memories of the physical world always lie in the past. To distinguish time, we have created three tenses: present, past, future. Our brains cannot comprehend another tense that could perhaps be between future and present, or present and past. Change in space is time, thus all that is present is to be past, and all that is future is to be present, but what about the past? What happens to the past moment? That is where memories come into place. Memories allow us to revisit

Concept of Memory

the past and recollect information that is not accessible in the present. Memories are an excellent mechanism for evolution as they allow the human to rationally think, remember, and act only according to the beneficiary results of past activities. Without memories, a man wouldn't be able to do anything, let alone survive. That is why memory is so crucial to understand. But there lies a deep secret about memory. The secret is that all memories are made up.

The concept of the past is actually not real. All that you think has happened to you is only an interpretation of what truly happened in the objective world. You have only remembered certain bits of that reality, and when you try to remember it now you try to join those bits to complete the puzzle. It is like dreaming. All your life you might be dreaming, and the moment you are living right now could be the moment of awakening. You would think that I am a certain age, and I have been with such a family for a long time, therefore I have lived in this world for a long time. Incorrect. You believing that your interpretation of the past in the present moment is the same as the exact past moment is flawed. Simply because you just cannot live the past moment again, you can only remember its fuzzy details. Even then, those details are highly examined and all connected by the brain to present you some sort of stimulus or image in the brain to tell you that the past moment was like this, and right now, it is different. Furthermore, the past is only interpreted in hindsight. It is interpreted in a sense that allows you to confirm your present moment. In this way, the past may have never occurred, the future may not occur at all. These defence mechanisms of your brain to distinguish space and time might be all fake. But this is all conspiracy, let's get back to what we know. We know that the past moment happened, but we don't know if it was real or not or if we can remember it correctly or not. But surely it had to happen right? It is true that the past moment did indeed happen but it happened in another world. My theory of world jumping is that we constantly jump from one world to another, and experience that world's objective reality. Each world is only one moment of our conscious world. And we cannot go back to the same world we went to, hence we cannot experience a past moment. We can get close to experiencing the past moment again, but we can never truly experience the past moment again because as soon as we left that moment, our understanding of that past moment changed. And if we think we are experiencing the past moment again, clearly that means though our interpretive past moment has matched with this current moment, it may all be just a fake interpretation; a fake judgement of their similarity. The bottom line is: we cannot experience the past moment as it was again, not even if we tried to recreate it.

It gets even more confusing here. The theory of world jumping suggests that we jump worlds, but what about the time between jumps. What happens when we aren't in a world, but between two worlds. This is when the human creates a new world, a new moment, a new life and a new world. Indeed, the human creates its own worlds, and jumps between them nonstop. Once he has created a world, he can no longer create it again, as that spontaneous action of creation cannot be replicated again. Thus between moments, the human plans, executes and creates new moments. This is the difference between thinking of reality and living reality. Before you live, you create that moment in your mind, and then you live in it. Once you finish living that moment, you imprint the moment's environment not its creation and continue to create new moments and live in those new environments. Such life of moving from one moment to another leads the human to believe that he has left something or that he is about to leave something. All this may seem bizarre and absurd, but it portrays a simple aspect of our life. That concept is that the past moment is an illusion. It is only an interpretation.

If the past doesn't exist, what about the future? Does the future exist? Nope. All thoughts and expectations of the future derive from the past. All moments that you would not be able to do right now or in the past soon become goals for the future. This tendency of believing that all will go well and I will be able to do greater things in the future is only a lie to yourself. After all, we will soon forget about our goals when we hop onto a new world, and only remember a glimpse of the

Concept of Memory

goals we created. In fact the moment you start thinking of the future, you start transitioning into a new world. Even a thought of the future creates a new world potential. You believe that one day you will have the potential to get to a place that you desire. Thoughts after thoughts entire your mind each creating an alternate platform of reality or worlds as I call them. Just for fun, you imagine yourself in those moments as well. This is a mockery for yourself to believe that soon you will be able to replicate this imagination to reality. This endless cycle of remembering illusions, creating new expectations, forgetting these expectations but believing in remembering your expectations ultimately leads to death. One moment after another, we constantly force ourselves to think of the leftover past, remember what we had planned to do so that we could do it, and then we dream about those plans in the future, telling ourselves that one day you will achieve what you had desired. Foolish. It just keeps you in this cycle of trying to remember and displacing some other expectation onto the future. Never are you truly able to achieve any goals, yet you are only able to deceive yourself into thinking of the goals, imagining those goals being achieved, and then moving onto new goals. Let me give you an example. A man wants to live in a rich house, and he believes that all this life he has lived in a very poor and small apartment. He remembers sharing a tiny room with his roommates and not having the privacy of doing what he wants with the space he has. The man then shifts from remembering that moment to imagining a brand new house, full of luxurious and very spacious. It gives him much pleasure and joy to think of being in this situation. He thinks about this house a little more. He thinks about what this new house will have for him and how it will be nothing like the old house he had. Soon this imagination stops and the man continues on with his day. The next day, the man remembers that he wished for a house, but does not remember what things he wanted in that house. In another week, the man remembers he wanted a better house, but is not sure if it was a house, or a car or something else. Another year later, the man thinks that all this time he had wanted a new car so he decides to buy a new car. The idea of the house had completely vanished from him simply because the man's own mind had altered the man's thoughts. In other words, the future that the man had created was only an illusion, and the past of having a small house was also an illusion simply because it was exaggerated. The brain likes to fantasize in the future simply to escape the mundane present. Though the man had a good enough house, in his eyes it was small. Even if he were to get a new house, he would forget the past moment of living in a small house then moving to a big house. All he would remember is that the new house is somehow annoyingly small or not spacious. In such a manner, the man would never be satisfied because his goal of having a larger house will never be met as the man will forget the transition from a small house to a large house; all the man will remember is that he moved, but he moved in a house he did not want to move in.

Such stories remind us that our created expectations and our memories of how our lives were are very fuzzy. Our brains do not give us a proper understanding of practicality, they only present options that supposedly make us feel good about ourselves. We feel good about ourselves when we remember we did a good thing in the past, so that memory is recollected more often. And as we know from the way memories work, if memory is recollected more often, it becomes more easily convincing. On the other hand, we do not want to feel bad about ourselves so we create defense mechanisms against our poor behavior. We give ourselves to reason and reduce that cognitive dissonance of the feelings of guilt and shame. In this manner, these memories are lowered and inhibited and soon left in the fog. Sometimes, these memories could be predominant and could lead to some psychotic disorders. Nevertheless, they are uncommon in general.

The past is the past, it has been gone, and remembering it and replicating it is only a fool's game. The future is only based on the past, which itself is remembered incorrectly. The future is based on nonpractical things out of our reach. We oftentimes don't want to believe in nonauspicious or negative things thus we exaggerate our thoughts. The future itself is also another illusion. What remains then is the present moment. Though we are hopping from one world to another, from one moment to another, we can do so very consciously and thoughtfully. Instead of jumping from one moment to another in search of an imaginary world, we could jump with recognition. We could truly experience this exact moment, and not think about the past or the future. In this way, our jumps will be less frequent, and at some point, we will stop jumping. We will be content with any world we come across, thus the jumping will soon lead to a smooth transition. No longer will our lives be in discrete memories or events. Rather they will be a continuous flow of awareness. We won't expect anything as the future is not in hands of anyone. We won't get caught up in our own biases as those biases are our deceptions. In this way, one would live experiencing, and not jumping.

The unknown future and the imaginary past are integral parts of our life because they appear to give us motivation and ambition. But on the deathbed, they haunt us and remind us that all our lives we forgot to live, we forgot to experience. All we did was jump. We just jumped from one event to another, and took no time in stopping our thoughts of the future and living in the present. We never lived, we just dreamt of living in a better place, but that better place was no better than our present place thus we kept searching. Goals and ambitions are a good thing, don't get me wrong, but if we are always searching for a better place, we will never be satisfied in any place because there is always a better place than the better place we had thought of originally.

Memory and Religion

Our knowledge of reality lies in our memory of personal experience and perceptive knowledge that is meant to be learned in school. Memory holds all these in one place, allowing things to flow coherently and making sure the stream of consciousness is able to flow without having barriers. If such barriers were to come into existence, such as a new phenomena or a new experience, our memory soon builds its image, its platform and its meaning to recollect of it in the future. In this manner, the river of consciousness is efficient in understanding the world, the scientific and perceptive world that is. We are what we think, and we think of only that which we know. So then the question becomes does our memory allow us to understand everything in the world? Our memory is knowing about ourselves, but what about knowing about the world around us? Our memory is categorized through personal experience and emotion, thus it is easier to know yourself. For example, all the characteristics you believe you have come from a stream of memory in the past that is recollected at a very rapid speed and thus allowing you to believe you are that which you think you are. There are two types of memories: long term and short term. Your most long term memory is probably associated with a really strong emotion of either excitement or guilt. It becomes evident that short term memory is usually the one that is used day to day activities, while long term memory is used for specific and conscious activities. Let's start by thinking about who we believe we are in terms of long term memory.

How far back can we go? We have evolutionary and biological predispositions as well as psychological and environmental predispositions that make up our memory. For example, all of our instincts and impulses come from the development of the sympathetic nervous system in the brain. Also known as the fight or flight system, the sympathetic nervous system allows us to feel a sudden burst of energy, high heart rate, and fa, st blood pressure under stressful and fearful moments. Though for our ancestors this brain mechanism helped us face fears, it is no longer needed to that extent in today's life. Yet, we still have it within us. A stick appears as a snake only because our brains are built precariously thus taking even the little and innocent things as potentially dangerous. That is only the sympathetic nervous system though, all of our proteins that come from our DNA have a long history. Our biological systems work the way they do because there is a biological memory of coding for a certain trait. As far as biology and chemistry go, our body systems such as respiration, digestion, cardiovascular system, and the endocrine system all have special anatomy each with intricate details of how they work. Such a level of intricacy is not regulated by conscious thought; we do not need to think to digest or urinate or so forth. These mechanisms have perhaps what people call evolutionary memory. Since our ancestors have been using such mechanisms for so long, these primitive systems have become independent from our conscious thought. The brain regulates the body and maintains homeostasis, but it does not command each organ system to work in a specific way. Each system works according to its phenomenology. We can go even further and say each cell, each DNA strand has a memory. When I say memory, it need not be sensual memory, I am talking about remembering the memory. How do the simple four bases(A, T, C, G) know what proteins to code for, what proteins to not code for, what systems to promote, and what systems do not promote? It becomes evident that science has only gone so far to say that we are what we are because of DNA coding. Yet, we do not know how DNA knows what to do. We know the intricate system of DNA replication, translation, and transcription but we do not know what system or intelligence promotes all of this. What I mean by that is how a biological phenomenon happens at such a peculiar and intricate level without any command from the brain, or any intelligence. Is it God who creates, sustains, and destroys all this? Is it evolutionary biology and just the fact that biological events happen for the sake of happening? Or is it the fact that there lies something further than our intellect, perhaps a metaphysical soul that constitutes directions for all cells in our body? Who knows, but I have a general idea, rather a theory on what is happening.

When using our intellect, there are three tasks our memory does. First, it recognizes a phenomena that appears in front of us, then it labels it/categorizes it according to our subjective experience, and at last it stores its necessary information for later use. Remember that just like language is acquired, our memory is also acquired. When we label and recognize things, we put forth boundaries upon them. This is the concept of differentiation. By having a linguistic structure and a fast intellect to associate and understand our physical world, we start remembering, knowing and finding out new things. This explains the short term memory, but what about our evolutionary memory? This is where we must understand that our intellect is limited. Our memory is predisposed to labeling and recognizing objects and phenomena according to our language, emotion and culture. Any memory is associated with a personal experience thus having a bias against the reality of that experience. But not just a bias, but the memory also shrinks the experience. An experience of being in a new place is completely lucid and clear. Yet when that same experience is recollected later on, we try to use words to remind yourself of what that experience was. Unless that same experience does not come up again, we simply cannot rely on our own linguistic mechanisms nor our memory's small bit sized experience recollection to truly understand that experience.

    Our brains can understand that which can be sensible, that which can be deduced and that which can be imagined. Imagination is still the cultivation of memory, I would say. Dreams are a perfect example of this. Some argue that dreams are transcendental and gateways for the soul to meet God. While others argue dreams are just unfathomable and incoherent ideas of our past, all collected in a disorganized way and displaced before our eyes. I would dream are unorganized and incoherent and that we try to make unnecessary sense of them when we wake up. But this incoherence is not in the specific dream content, but rather in the content's context and connection with other dream content. The images or vivid events in dreams come from our own daily life and even the activity happening in the dream is a product of our memory, but the connection of one activity to another is something that our memory does not provide for. We might be dreaming of being on top of a mountain, which we had visited a long time back, but only this time, suddenly we see lions rampaging behind us as we try to save our lives. Though this exact event is not recollected in memory, the conjunction of two different events creates a completely new experience. As Freud argued, since we do not have conscience or a sense of morality in our dreams, we are prone to doing things we would otherwise not do when we are awake.

But my point of telling you all this is that imagination and dream is still yet just another job of our memory. Just like our thoughts, language, feelings and religious beliefs, imagination is also just a cloud of information which is accessed while either sleeping or daydreaming.

Transcendence of consciousness is the hot topic of this generation. Transcending your predispositions is a very daunting task, nonetheless, it has been done throughout history. Many religions have accounts of people transcending their bodies and brain. And to those nihilists and skepticists, I dare to say that religion may seem like an old woman's tale or a confabulated story of some sort, but with enough practice, I believe religion can allow us to transcend our predispositions. All religions argue on the concept of a divine being that creates life, thus they believe they must follow the teachings of the God written down in their respective scriptures. Herein there is a subtle message which is that God will take care of us if we do as he says. When it comes to metaphysics, religion is the first one to jump on the boat. But let's leave the specifics aside, the bottom line is that our physical bodies can be transcended to our true and pure selves also known as souls or spirits. We cannot fathom this nor transcend it in any manner while using our intellect and memory. Our memory and intellect are useful for our physical lives which promote evolutionary biology. Yet beyond this biology lies an even subtler level of mechanism that goes beyond the physical self. Whether you believe you have a soul or not does matter, because beliefs are still creations of our culture and memory. I think religious scriptures themselves portray no real value, rather it is the guidance or philosophy of those practices that have value. Reading, writing, watching, or even researching about God or our souls while never getting you any closer to an understanding about them. It is through a monastic and meditative life, that you will be able to transcend your predisposed physical selves. Transcendental meditation is hard to describe through words, but simply put it is the practice of being mindful of our breath and body. In my opinion, this is a better approach to trying to understand our true selves. The scriptures guide us to live a moral and ethical life, and they even give assurance of going to heaven or hell depending on our actions. Yet the scriptures don't give exact assurance that all that is in the scripture is true. One cannot blindly believe in religion until he experiences the religion's impact on his personal life. Many religious people have firm faith in their religion and say that they have ever experienced an encounter with God himself in their dreams. Near-death experiences are also associated with feeling as if reaching heaven. But the bottom line is, that everyone knows themself and they know whether they are simply believing something or whether they are acting upon their experience. Meditation can provide this experience but don't listen to me or any other guru or a monk that tells you this. After all, e might be wrong and deluded in our ways. Rather, try it on your own, and if there lies a soul within, then that soul will slowly convince you that through meditation you can connect with your true self. If there is no soul within you, then at least meditation helped you cope with your psychological issues. I see no harm in meditating in a secular fashion. Zen Buddhists have no deity or God that they preach about or pray to. Rather they believe that the Buddha was an ordinary man and that all he did was ride his predispositions through meditation. The question of soul, existence, and afterlifee is not important for these Zen monks. It is all about being aware of our predispositions and lowering them one by one. The analogy is that our pots are filled with lots of water at the bottom, yet the weeds and grass are the only visible things on top. To reach our 'true selves, we must first get rid of our hindrances or unnecessary predispositions that appear on the superficial level.

    So God need not be real or fake, but one sure thing is that our memory, short term, long term, and evolutionary, inhibits us from realizing our true self. I postulate that the reason why babies are so happy and make others happy as well is because they haven't developed any memory. Though they have evolutionary predispositions, their body is still in the process of understanding them. I would argue babies, unknowingly, realize their true self. Yet as the baby grows, he soon builds up his psychological and behavioral predispositions, only to pile up on his previous evolutionary dispositions. Our memory helps us survive, love, and care for one another, but it also inhibits us from looking at ourselves at a broader level- a metaphysical level.

Early on, I posed the question about how our biology works at such a magnificent level. Well, the answer to that might lie in understanding our metaphysical self. Though it might seem as a nonexistent concept, we cannot be sure of anything that we have not yet experienced. Balance is key. If we become too ignorant in thinking that there lies no metaphysical life, we might lose a great opportunity of using this life to liberate. On the other hand if we become engulfed in blind faith, we would simply keep lying to ourselves of something that does not exist. Yet the middle way, the balanced way suggests that one must accept both philosophies and follow one or the other after trying out both of them. If experience suggests that there is something metaphysical within us, then one can go onto pursue it. If experience suggests that there is no metaphysical self, one can go back to his daily life, but not preach to others according to his experience. Everyone should experience meditation first, then only make conclusions about it. Maybe it was our ignorance all along that kept us away from realizing ourselves or maybe we were right all along about there being nothing but our physical bodies. The only way to find out is introspection through meditation.

Dreams

Many people say that our lives are like dreams. Maybe death is the ultimate wake-up call for some people, who knows. But I argue that our life is not a dream. It is not a dream because it is filled with awareness. Our ability to be focused and concentrate on one thing suggests that our life is more than a dream. Surely our life would be a dream if we never focused and constantly thought of the past or the future, but when we leave those aside and live in the moment we find that life is completely different than a dream. When I say live in the moment, I mean experience every moment of consciousness. For example, when I am eating an apple, I am consciously noticing its red color, its juicy taste, its texture, and so forth. Not a single time, am I associating the object or the apple with another experience from my past. Because when we start associating with experiences, we start losing real lies and fall into an illusion. Dreams are quite the opposite of the present life. We dream when we are unconscious, not when we are awake. We dream about things that make no sense or that have no meaning in themselves. We dream about phenomena that are shallow and unknown. Freud's book, Interpretation of Dreams, gives an excellent view of how dreams are in nature unorganized and absurd and can never be understood while we are awake. Many authors after Freud tackled the concept of dreams, and in today's era, we have come to a conclusive conclusion that electrical activity in the pre-frontal cortex and rapid eye movement(REM) are factors of dreaming. We still don't know why dreams take place and why they occur during sleep and not wakefulness. Nonetheless, dreams are a great way to reorganize and restructure your memory. It is proven that insomniacs have on average less memory than noninsomniacs. The reason for this is that deprivation of sleep leads to lower long-term memory.

    In 1953, H.M had his hippocampus surgically removed due to excessive seizures and skull damage. H.M went on to live a life of no memories. The life that no one could dream of. He could not remember anything he did, sensed, or thought of. He was a man in the present moment. He watched the same shows, played the same games, and did the same things again and again and never felt bored. In fact, every time he did something, he thought he was doing it for the first time, and thus this exuberant feeling of trying something new every moment gave H.M some type of relief, or so the doctors believe. The reason why I shared this study is because it shows that damage to certain parts of the brain can lead to memory loss, especially long term memory loss. Who knows, maybe the life H.M lived was much more exciting and fruitful than if he had lived a normal life. Maybe he was more content in being able to not think all the time, not having to worry due to having expectations or horrible memories. Every case has its pros and cons.

But how did this case relate to dreams. For H.M, life did seem like a dream. He believed that he had just woken up from a dream and was experiencing his present moment for the first time. A life that felt like a dream is certainly very scary, but is it really that bad? Is it only scary because we haven't experienced it or is it scary because we like to hold onto our memories. I would argue that not having any memories(besides the primitive memory of eating, drinking, sleeping and so forth), would be better than having both good and bad dreams. Though, I believe that this speculation is fallible as there is no experience behind my claim. So the closest to that we can get to H.M's is attenuating our past thoughts, and concentrating on our present moment. Not necessarily waking up from a dream, but I would say gradually transitioning without looking back too often.

The memory of this chapter

This chapter was very hard to write about, and some of the concepts are still puzzling to me and my friends. The theory of world jumping portrays a key concept of our existence: the drive for change. We can't stay stagnant and we want to move on, thus we constantly create a present, past, and future. If every day was the same day, hypothetically, we would soon rather die than live a mundane and predictable life. We need thrill, action, and unpredictability. If we know what is going to happen, then life becomes unseemingly boring. But on the other side, if we continuously look for unpredictability and constantly create expectations for the future, then we are putting our life in hands of plain luck. We are mortal beings, and we will die one day. Our goal in life is not to create memories. It isn't to create goals and satisfy those only to find there are more goals to fulfill. It is to experience, explore and enhance this exact moment of awareness. And the question about our souls, well that is just another concept that lies outside of our linguistic methods. As they say, shut up, stop thinking and sit down and meditate. This moment of life is life, and beyond that is all just speculation of the brain and mind. If there is one thing to be remembered from this chapter, it is this line: Live life for the sake of living, exist for the sake of existing, experience for the sake of experiencing, for once that moment is gone, it will never come back, not even in memories and not even if you try to recreate it in any shape or form.

11. Concept of God and Free Will

Social Self Concept

All knowledge boils down to this phenomenon: the concept of Free Will. Are humans ever capable of understanding what lies beyond their capacity? Do they have any say in what is predisposed on them? It becomes evident that all philosophers have pondered upon human will and life and death. All philosophy has questioned the concept of self. What does it mean to be me? How do I change myself? Can I change myself? Do I have the ability to be in my own control? What does a person that is in their control look like? Why would I even try to figure out my true self? Such questions will be examined in this chapter of the book. But it all boils down to one big question: Who am I?

Psychologists have referred to our understanding of ourselves as self concept. Self concept is knowing yourself by extrinsically examining yourself. This means you refer to yourself according to your beliefs, culture, religion, nationality, gender, and so forth. These are all extrinsic factors as they are created by the outside forces. People look at you in a certain way and you believe yourself to be that way. The keyword is belief. Believing something and knowing something are two different things. You know you are a male because of your genitalia or you know you are an African American because of the color of your skin. But you believe that these concepts have a particular meaning. These factors of race, gender, ethnicity, and so forth have a predisposed notion to them. This means there is a stigma that lies within the society toward these factors. For example, a tall black individual, no matter how kind and warm-hearted, will be approached with caution. A short Asian lady, no matter how violent and ruthless, will be approached as if she is innocent and kind. None of this is justified and in fact, this is a bad habit of our own psychology. The point is factors of physical nature have a deep-rooted connotation in today's society. We make sharp judgments about people just by characterizing their physical appearance. And you might not even know you are doing any of this judgment. At a social level, today's society is very incompetent at being open-minded and with less judgment. I think this isn't a generational problem. This is a psychological problem of self and others. I have already mentioned how the environment and surroundings can condition someone to the point where they become puppets being played by societal norms. But there lies an even great trap within all this. That trap is a delusion about free will.

Humans do actions that emphasize their self-concept. We are always seeking society's approval, and once we get this approval, we try to strengthen it even more. The norms that have been conditioned upon us are very deep-rooted. Just the last chapter, I mentioned evolutionary memory. I might say I do an action, but what is the I? Am I the genes that code for this behavior? Am I the environment that shapes and conditions my behavior? Am I the food I eat that then is converted biochemically into proteins and carbs? Or am I neither of all these? Am I just a product of all these? Is there even a singular entity that creates, sustains, and destroys my self-concept, or is it multiple events, people, and processes that result in my action? And is it even possible to know my real self even if there might be one? A drunk man never realizes his condition until he is sober, nor does a human realize he was in a deep sleep until he is awake. Similarly, if all these predispositions are simply illusions, then how is it that I can know my true self without clearing them out first? Trying to find answers whilst having such predispositions is simply creating more illusions for oneself. There is no truth in them, it is just their relevance and connection to our logical and axiomatic understanding of the world. All our understanding of the world and ourselves comes from a singular and personal perspective. This perspective is collected, recognized, and stored by our brains. Our memory holds onto categories for further analysis. Aren't we thus predisposed by our singularity? If our ego is in control, then our experience also becomes narrow and highly subjective. No two humans act in the same way in various situations, no matter how well they are conditioned and how similar their genes are. Though our psychological dispositions do play a major role in our behavior, there seems to be a little more to the story.

Up until now, the scientific community has preached to us to advance technologies to advance our understanding of the world and us.is helpful as technologies have lesser dispositions than humans. A human is bound by emotion, experience, morality, and personal bias and in such cases, he might not act rationally or appropriately according to the situation. In such cases, a robot will help us close gaps. Things that humans can't do, or don't want to do can be handed off to artificial intelligence. But to what extent can robots fulfill human wishes and close the gap in their intellectual knowledge? After all, robots are creations of the human brain, aren't they?

We have attempted to use our brains to create something that transcends our brains. This has been successful to a certain extent. Calculators give break time to our brains, showers and home facilities give break time to our brains, and transportation, and communication through fast facilities gives break to our brains. We have made our life efficient to give a break to our brains. Thus, making our life more efficient, allows us to use our time not to worry about our survival necessities, but to use that time for either entertainment or advancement in such facilities. A modern man, no matter how rich, does two things in his free time: one is that he spends time with his family and entertains himself through provided facilities, and the second is that he somehow advances the facilities he has been given. For example, if he is a doctor, his work is helping treat patients and doing research for curing new illnesses, and his compensation for this is that he gets a salary with which he produces personal entertainment. In a capitalistic system, this is enhanced even further. The system's job is to continue to provide facilities and goods and services for the sake of advancing the system. Children and adults are being brainwashed day and night by television and advertisements that lure them into this system. The quench for wanting more seems to never end as the system keeps bringing out new products and new services for society. We pair such political systems with advancements in technology and we have the United States. The economy in the United States is controlled by two fields which are the healthcare service providers and the technical services, providers. The latter is seeming to only grow, as there is much more in the field to advance into and the fore is still dominating yet it is nowhere near the rate at which the technologies are advancing. The robust companies in the US include Facebook, Amazon, Alphabet(Google), Apple, and Samsung. It makes sense that these two industries are the ones that are growing. When we are not worried about our health or body, we are usually engrossed in social media or some other type of technological service. Noam Chomsky, a contemporary linguist, argues that such large media usage has a direct impact on our ideology and will. The environment in which we grow is heavily reliant on our phones, laptops, the internet, and other electricities. It has gotten to the point that it is impossible to survive without having a connection to the Internet. Just a couple of decades back, electricity was found, and now it just transformed into the Internet, which is by far the most time-consuming service of today's life. The Internet and the media have only created a strong impact on our thought and knowledge systems. So now I question how can we control ourselves when the media is constantly changing our beliefs, viewpoints, habits,, and other feelings. The answer is we cannot. We cannot free ourselves from the Internet because we cannot free ourselves from our modern daily lives. Just scrolling down Twitter or Instagram, we absorb so much irrelevant but eye-opening information that changes our entire mood and thus changes the rest of the day. Watching a movie or two completely draws us into an alternate reality that was simply directed and created by humans. The day-to-day activities require us to go through advertisements, whether that be consciously such as social media or news, or unconsciously such as pop-up ads on websites. The point is that though we believe we constitute our life and we control our life, there are multiple factors, including biological, evolutionary, psychological, and at last social ones that constitute our being.

All such factors are hindrances and predispositions to our true selves. The social and psychological factors are man-made and they affect our biological and evolutionary factors. What we do with our behavior and environment affects natural selection. And neuroplasticity has already suggested that our brains can change and will change if they are challenged by environmental events. Though we can't change our evolutionary dispositions, we can certainly slowly use them to our advantage to live a healthy and happy life.

Religion

Lovely is the word religion, lovely is the act of praying, peaceful is the goal of devotion and joyful is the attainment of heaven. Religion comes in many sectors and it is very closely connected to modern-day riots and warfares. But leaving aside these poor results of religion, we must dive into the concept of religion itself. What is religion and how does it solve the self-concept problem? Is it just a defense mechanism, a social construct, or a genuine solution to our sufferings? What does death mean and how is God related to it? In fact who is God and how does he choose what happens to me? Is there a metaphysical soul that meets God after death or is that just a tale told to stop wasting our time thinking too much about our mortal selves? Whether God exists or not becomes a second question, but the greater question is why does the concept of God matter so much.

God is generally defined as Omnipotent, Omnipresent, and omniscient. Omnipotent means he can do anything and everything possible. Omnipresent means he is present everywhere the human can travel or could travel if he were given the right abilities. omniscient meaning he knows everything period. All that there is in this universe, and the next one is determined and known by God. Beyond this orthodox concept of God, there are many other concepts of God. In Hinduism, gods are represented as forms of energy. It is believed that all humans come from that energy and go back to that energy when they have been liberated from the cycle of life and death. In the Abrahamic religions, it is believed that if one were to do good deeds and follow God's doctrine(respective religion's scriptures) then he would reach heaven afterlife. In more philosophical religions, God is seen as a concept, rather than a motivating tool for humans to do good and leave evil. These religions include Buddhism, Jainism, Confucianism, Taoism, and Shintoism. God might be a concept, energy, a living being, or something we can never imagine but it, certainly,d promotes well-being and doing good. All religions premise their philosophy on doing good. From a logical perspective, one can say that humans have created a system where they can live safely and healthily by believing in God. While on the less logically driven side some believe that this world is all an illusion created by our senses. In Hindu and Buddhist religions, the main emphasis is on the exercise of detaching from our worldly desires. This is predominant in Jainism as well as it emerged before Buddhism and helped the first Buddha, Siddhartha Gautama, attain enlightenment. Detaching from desires is not a sign of rebellion, but rather a sign of expressing less interest in this physical world. Monks of all sorts become ascetic because they believe there lies a metaphysical universe such as the state of Nirvana, Moksha, or some sort of enlightenment that can be achieved during this lifetime. It is important to note that reaching such metaphysical states of mind is different than reaching heaven. Heaven and Hell are only reached after death while these other states of experience can be experienced during life as well. Buddhism is very close to our scientific and logical reasoning understanding so let's take a look at what it preaches.

Buddhism and Science

Secular Buddhism is very interconnected with modern-day science and psychology. Siddhartha Gautam, before becoming a Buddha, was himself an atheist. Though born in a Hindu family, he questioned its traditions, regulations, hierarchical structure(caste system), and morals. This is exactly what science is doing today. The scientific community is attempting to get rid of any dogmatic and abstract concepts and trying to understand them through specific and scientific methods. Many psychologists, behavioral therapists, and psychologists have already implemented meditative practices in their work field. It becomes evident that Buddhism is more so of psychological treatment or philosophy of thought rather than a religion.

The foundation of Buddhism is on the concept of suffering. The four noble truths are that suffering is inevitable; there is an end to suffering; suffering is caused by attachment to impermanent reality; to end suffering completely, one must follow the eightfold path that the Buddha himself laid out for the generations to come. Buddhism wasn't founded to reach heaven or hell, nor did it preach about a divine being or any other deity that had control over human life. Buddhism taught that the human mind is the greatest controller of the human and that if the human wanted he could control his mind and rid of all his suffering. Buddhism comes from an existential point of life and death. The Buddha himself felt depressed that one day he too shall become old, ill, and die. No matter how rich or poor one may be, no matter how smart or dumb one may be, they all shall die one day. This sense of impermanence lies at the core of Buddhism. It becomes so important that to reach the final stages of enlightenment, one must escape from all his psychological and physical desires that hinder his true existence. All that has formed is unreal, all that appears will have to disappear, and whatever arises in the mind will have to cease one day. Such thoughts then give us the understanding that nothing in life is permanent and one mustn't cling onto his possessions or desires so much that he soon feels incomplete without them. Our thought gives rise to reality, and this reality then further creates more thoughts which then continue the cycle on and on. In this way, we forget that the real moment is the present moment that is undisturbed by our emotions and feelings from the past and expectations and desires of the future. The Buddha taught us to be equanimous in all situations because all situations arise and fall and have no meaning, rather it is the mind that creates their meaning and associates them as good and bad, necessary and not necessary. These extremes are the ultimate illusions according to the Buddha. We must be balanced regardless of the external factors that come and go because these external factors have no relation to our internal wellbeing. In this manner, Buddhism tries to reduce anxiety, stress, and suffering by preaching about living a virtuous, balanced, simple,e and calm life which is focused on getting rid of our predispositions that arise from our attachment to worldly possessions and desires.

All religions come from an existential foundation. Why do we exist? Who made us? Who made the universe? Why am I here? What is the point of life if I am to die one day? Such questions are easily answered by the religious doctrines as they ensure people that the purpose of living is to attain a certain state of existence. Whether that be afterlife heaven or nirvana or any other state in which our existence overcomes its death. Buddhism approaches it in a little different manner. Though nirvana, the separation from our body and brain, is the ultimate goal of Buddhism, there is more emphasis on the practice of reaching nirvana. In Buddhism, there is a saying that 'I am already home.' This means that the process of meditation and mindfulness itself is the goal of that practice. In other words, virtue and holiness are themselves rewards for the practitioner. One might ask how does this make sense? Well, the concept is that when one does good, one *feels* good and when one does bad, one *feels* bad. Good and bad are still concepts in our minds, so how do we know if we are doing good or bad. To this, the Buddha says if any action is done selfishly and with a motivation to harm someone else then it is bad. And if an action is done selflessly and with a motivation to not harm yourself or anyone else then it is good. This seems to be the axiomatic belief of all Buddhists. Zen Buddhists do question such topics and create riddles out of them called koans. These koans suggest the absurdity of language and how the practice of Buddhism isn't about following an ideology or a principle, rather it is based on mindfulness and the practice of being balanced and selfless in all situations.

According to the Buddhists, we do have free will. Every person can know their true self if they uncover their predispositions and hindrances that are brought upon them by society and their biology. To do this, one must tame his mind which tries to run in all directions. They call this the monkey mind, which goes from one event to another, one thought to another, one person to another, one idea to another, and so forth. If we can realize that all thoughts, actions, feelings, emotions, and desires are momentary, then we can renounce them and live a simple and mindful life. When there is less for the mind to worry about, it naturally starts to become calm and peaceful. We do not have free will until we are controlled by our brains. It is believed that this brain and body are impermanent themselves and thus they do not resemble our true entity. This is where we dive into the realm of metaphysics. Buddhists believe that our existence is eternal and that we are not bound to any particular physical body or object. We were never born, and thus we will never die. It is only the brain that was created and will be destroyed. Our memory of experience and thoughts will cease, but our true self will not because it was never there, to begin with. We simply create a narrative for ourselves to create a subjective and personal experience. We give ourselves a name, an identity, and soon believe that this momentary identity is our real self. But the Buddha said that with his own experience he was able to realize that there is something beyond this physical body. If one were to slowly train his mind and loosen his connections between the mind and body, soon he will realize that the body is just a vessel for his existence. This then springs the idea of rebirth and how our soul moves on from one body to another depending on our good and bad deeds. No one can argue whether rebirth is real or not until they have experienced it for themself. For this reason, the Buddha himself states that one should not follow someone simply because of blind faith or family heritage. Rather, one should try things for himself and find his path/answer to his existential crisis. The Buddha was just like many of those who do not understand the purpose of life. But he didn't give up on his life. He meditated day and night without doing anything else. He took two meals a day from nearby villages and survived on minimum water. He lived alone and talked with very few people. As you can see, he was dedicated to finding out an answer to his question. He was so dedicated that at some point he gave up food and water. Later on, he would realize that such drastic means of practice are of no use. Because all such practices come from extreme philosophies which do not have a balanced mindset. When our practice is balanced, we can reach further our spiritual goals. Thus soon he started eating, drinking, and lowering his extreme penance. After months of meditation under balanced penance, he soon reached enlightenment. Siddhartha was no longer the identity of this being as he no longer identified with himself. Rather, he was called the Buddha because he had arisen over his intellect and mind. Bu means intellect and dha means above in Sanskrit and thus his identity was him who had risen over his intellect and physical self.

The transformation of the Buddha took a long process. In total it took 6 years of penance and hard work for him to realize the meaning of his true entity. But this knowledge that he received after so much hardship was easily given to others by his teachers. What took his decades to understand, he told others in just a few hours. The only difference was that Buddha had experienced his philosophy of nonself. Those around him had to yet realize that and so language alone wasn't enough to get them to enlightenment. The same applies to all of us and many Buddhist monks. Those who follow words and the philosophy of the Buddha simply as intellectual wisdom never reach nirvana. It is only those who experience such wisdom during their meditation sessions that attain nirvana.

What we believe ourselves to be is not our real selves. As I mentioned previously, the self-concept that we believe we consist of is only a creation of society and our impermanent brain. When we realize that the nonself part of us goes beyond any language or philosophy, then we truly understand the teachings of the Buddha. It takes many years to understand this, let alone put it into regular practice. But once we have experienced the state of non-self, we realize that free will is not just the ability to control this single body's actions and thoughts but it is the ability to control the entire nature. When the Buddha had taught all that we could to his disciples, he brought a lotus and told them that the remaining wisdom he had was no longer able to be shared through language. Thus the symbol of the lotus provided the means of sharing his wisdom. Only one out of all the disciples realized the meaning of the lotus and soon meditated on it to ultimately reach nirvana. The point of this story is that language can only take us so far. Our singular entity produced by our brains is limited and thus we mustn't always conclude that intellectual wisdom is the highest wisdom because intellectual and philosophical wisdom is in itself limited to its parameters of logic and reasoning.

Scriptures, deities, and rituals are all symbolic as they do not provide any literal meaning. In Buddhism, none of these aspects are present as it is strictly focused not on realizing God but on realizing one's true self. Maybe these two things are the same, who knows? Maybe the non-self that the Buddhas meditate for is the God that the religious people pray for. Nonetheless, the Buddha brings up a great point about knowledge and wisdom. He says that until one experiences the state of true liberation, that is the state of being nonself, one cannot end his suffering. Thus no one can preach to another about wisdom, no one can teach anything as it pertains to expression and language. All wisdom about our true self is to be derived from oneself in his method. Blindly following something is the last thing anyone should do. The truth already lies within them, whether they want to pursue it or not is up to their dedication and willpower.

The Cynic's Argument

The Buddha's path of enlightenment suggests a very optimistic and peaceful result. It showcases to us that all of us can end suffering and that we can find real meaning in our life. That we can live compassionately, selflessly, and without clinging to our egoistic singular identities. But this approach is argued against by the cynics as they believe that no one can transcend their ego. A cynic is someone who follows the philosophy of Diogenes, a Greek philosopher who lived in 400 BC. His philosophy is surprisingly similar to that of the Buddha's philosophy but it denies the transcendence and nonself aspects that the Buddha preached about.

Diogenes lived the majority of his life on the streets owning very few things. He was often called a dog because many dogs surrounded him at all times. Diogenes believed that one must live a life of virtue in agreement with nature. Thus he claimed that a man has to give up all his materialistic desires and live fully according to his nature. Diogenes had no shame in masturbating or even peeing in front of people as he believed that these actions are neither good nor bad, they are also perceived as bad because of social dynamics and passed down morals. Cynicism later advanced into a philosophy that promoted freedom. Freedom from society's dispositions as per the cynics. The dogs were a symbol of Diogenes' philosophy of freedom and happiness. The dogs lived a simple life and did not have any egos in themselves. They too urinated in public without feeling a sense of shame. They lived their lives freely with no psychological or societal barriers predisposed to them. As Diogenes died, his philosophy continued to spread across Greece and other areas of Europe. The word Cynic soon transferred in the Roman Empire where a cynic was someone who mocked and made fun of others as he believed everyone was selfish and no matter how virtuously they acted, they were still prone to their ego. This is the modern concept of being a cynic. Nietzche, another promoter of cynicism appreciated Diogenes and his effort to educate people about their predisposed concepts of morality and sanity. In Nietzche's eyes, there is no God nor is there a concept of morality. These are all created by human intellect to suppress their true nature. According to Diogenes and Nietzche, humans should live freely without suppressing their natural selves even if that meant giving up sanity and urinating and masturbating in the public.

Concept of God and Free Will

This approach to meaningless life seems very childlike and ignorant at first glance. But this is exactly what the Buddha taught. The Buddha also attempted to educate people about their mind and their ability to judge and create illusions about the past and future. Though he was on the same page with Nietzche on the topic of God and the concept of blind faith, the Buddha never advocated leaving our sanity. Though we should leave our possessions, wealth, and other desires we should in no manner cause harm or pain to others in the process of doing so. This is where Nietzche would argue that humans can never act altruistically because their sense of morality always revolves around themself. They act good because they want to be seen as good, and when they are in power because of their virtue they leave that virtuous life and become evil. Diogenes would add that the Buddha wasn't free because he was still bound by so-called 'good acts' or 'morality'. To be free, Diogenes would say that we must give up all our previous understanding of morality and strictly live in the present moment. The argument between the cynics and the Buddhists goes on forever. Nevertheless, they are trying to achieve the same thing: liberation from their previous self and their own physical body. Nonetheless, the cynics, in my opinion, follow the path of insanity while the Buddha follows the path of sanity. The cynics have not liberated from their evil self, the Buddha would argue. The cynics have only freed themself emotionally and physically from society, but there remains a sense of revenge and anger in a cynic which then keeps him away from enlightenment. Who knows whether cynics also get enlightened or not or whether they understand about non-self or not. I would argue that they would never really understand themselves until they give up their revenge. A cynic acts upon his revenge or anger, thus he is only reacting to people. While a monk acts upon a balanced mindset which at no time causes suffering for someone else. Both sides wish to liberate themselves from their predispositions, but one side does it in a balanced way and the other side does it responsively.

We are not selfish, it is only our mind that creates this feeling. We are not selfless, either, as that too is created by our mind. We are simply nothing, according to the Buddha. Until we give ourselves an identity, we have not liberated from ourselves. The moment we realize that we remain in between good and bad is the moment we become calm. Cynicism in that manner does a terrible job of being balanced. Cynics cannot maintain their emotions, let alone their philosophy of thought. It becomes clear that the monks win over the cynics because they can control themselves and remain balanced in all situations. However, the cynics do have a large hand on the topic of self-perseverance. It is a daunting task to give up shame and guilt and urinate in the public. It is a daunting task to continuously deny our morality and become inhumane. To some extent, the cynics have mastered their practice of being non-self. But they still have a long way from being enlightened, as they have yet to understand that they are still being held by their emotions and responsive philosophy. One thing that we can learn from both the cynics and the Buddhists is to question the concept of life, to question good and bad, and not blindly follow society by conforming to its status quo.

Nature as God

In both philosophies(cynicism and Buddhism) there lies a deep respect for nature and natural life. It seems that religion is heavily associated with nature. We see nature as pure and symbolic thus we have many botanical gardens around a church or a temple. The cynics saw nature as the ultimate phenomenon that transcends people and objects. Though humans will come and go, Mother Earth and nature will keep providing the same facility to all that come to it. There seems something interesting about nature that makes us feel so calm and peaceful. That is that nature does things as it is supposed to. There seems to be an objective reality for natural phenomena unlike the subjective reality experienced by humans. Trees, plants, sky, clouds, water, fire, and all other natural aspects that preceded human existence have no particular brain rather their essence lies in existence- the natural existence. Unlike trees, we have brains that acquire and produce knowledge and reality. The trees do not have a self-concept nor do they ever think they exist, well that is at least our thoughts on what they think. Who knows, the trees might be communicating with one another at a metaphysical level that is not apparent to humans. Many animals transcend human ability. That is bats can use echolocation to see their prey in the dark, platypuses can feel and distinguish water patterns to run away from predators and catch prey. The dogs have a peculiar sense of smell that allows them to 'smell their world- interpret their world through smell.' We as humans have a larger visual cortex in our brains and smaller auditory and sensory cortices suggesting that our world is highly based on visual cues. Reality perceived through one sense or the other completely changes the experience of the individual. We can never experience a dog's world, nor a bird's world simply because our brains are not capable of even knowing what such worlds look like.

We only talked about Earth, what about other planets? Other planet systems? Other milky ways? Other universes? It becomes a search for endless space. The more and more we try to examine the universe the more and more we realize our limitations. Nature transcends all beings and is the real god. Shinto Buddhism is completely revolved around nature. There are deities corresponding to natural elements, natural phenomena, and so forth. Shintoism, which is the indigenous religion in Japan, puts more emphasis on nature and its signals rather than God and his teachings. Signals from natural calamities and disasters are often signs of the deities being angry. On the other hand, calm weather and blossoming gardens are signs of prosperity and peace. It isn't just Shintoism that portrays this dedication toward nature. All religions have their ways of appreciating nature.

There are many folklore and religious stories that portray how important nature is. Being one with nature is often seen as being alive- being pure. This is simply because we accept our fatality and understand that what we are is all because of nature. Our creating, sustenance, and destruction are natural phenomena that do not require a species of some type of intelligent being to take decisions. Rather these things happen for the sake of happening. One cannot ask how the tree has enough intelligence to reflect wavelengths corresponding to the color green and absorb the energy that corresponds to other colors. One might say the genes have coded for proteins that have specialized jobs for each of the tasks required for the plant to survive. But then the question becomes how does the base-pairing have the intelligence to come together in such a proper fashion and make the right protein. Intelligence allows events in the universe to work in a certain manner. We come to quantum physics or theoretical physics which brings up another idea that our reality is distorted simply by our observation. Our prediction of certain events causes us to view them in that order as showcased in the double-slit experiment. Whether it be physics or biology, the universe works in a particular fashion. This transcends all intelligence as we are not able to replicate our universe. We can try to create microcosms of such ecosystems but never achieve such a large-scale and complicated universe that allows its inhabitants to question and be curious about their existence.

Perhaps we see this universe as God because after all, God is omnipresent and this universe, though a creation of God, has God within it. Causality is perhaps the issue here as we are always looking for cause and effect patterns. Physicists like Richard Feynman conclude the universe works as it is meant to work, it exists for the sake of existing, and thus we may not try to look for a supernatural entity governing all events in the universe.

There may be no God, and in my opinion, one mustn't believe in God simply because there is no other explanation for natural events. The way molecules collide and interact may be random and have nothing to do with a divine being. There might be no intelligence governing the nature and the universe that exists around us. God may be dead as well as Nietzche proposed. But skepticism leads to ignorance which then leads to unnecessary suffering. Existence precedes essence according to Jean-Paul Sartre, a famous existential philosopher. In some ways, this makes sense as meaning is only a creation for him who creates it. When there is no creator, there is no creation and thus our existence itself serves as meaning. Yet, there lies optimism in the most pessimistic heart. As Camus writes, we shouldn't depressingly look at our absurd life. There must be an understanding of our absurdity that will help reduce suffering. When we know that our lives are simply products of our thought and have little to no meaning in terms of the grand universe, we can free ourselves from unnecessary anxiety and stress. Positive nihilism is very helpful in some cases as it provides insight into our deeper selves. As Buddhism also mentions, there is no self nor is there a world that is around such a self. Space and time are creations of our intellect which help sustain life. Once again, I will repeat, when there is no birth, there cannot be death. When there is no physical entity, a real universal entity, we cannot be afraid of it ceasing, for it never exists in the first place.

The meaning of our lives and this universe cannot be comprehended through our intellect and so our only choice becomes to leave aside this problem, right? I think not. We cannot just give up finding meaning in the apparent meaningless world. All may be just an illusion, the self we proclaim to be might be just a fancy collection of beliefs and ideologies we have acquired in this mortal life, but what is clear is that there is still a lot to discover about oneself and knowledge. Intellectual knowledge about the universe may be not objective as we must look at different perspectives so that no biases come in the way. But the wisdom that perhaps comes from introspection and mindfulness might be a key to opening all of our existential problems. As the Buddha did, we may as well become enlightened in some way or another other. The knowledge systems can be learned through reading, writing, researching, or taking courses. Or on the other hand, it could be learned through introspective practices that our ancient philosophies suggest. You miss all shots you do not take so might as well try things, fail multiple times and arise from all to win in the end. We all will die one day or the other, the problem isn't that we will die, the problem is that we will feel the guilt of never truly living. All humans should strive to experience, explore and enhance their consciousness through whatever practice they seem to believe in.

12. Enlightenment

Summary

Throughout this book, I attempted to present my perspective on several concepts of our lives and the world. As always, the thoughts that arise from reading the book are more important than the mere words and content of the reading. We started by talking about language, and its fundamental error in communicating knowledge. Ironically though, I have done nothing but build on that error by facilitating more words. However, language is the only way we can share knowledge in this world so I had no other choice. After that, there was a discussion on knowledge and what knowledge means. We distinguished knowledge from the wisdom and said knowledge is deduced through the world, while wisdom is induced from within. Seeking knowledge is an instinct of all beings. Whether that knowledge is related to survival or quantum physics, all beings seek some of the other types of knowledge. We also distinguished between perception and sensation, which was greatly elaborated in the next section. The five senses give us the experience of reality and produce the perception of what lies out there. Furthermore, I gave an example of how differentiation is the major factor in all sense perception. All stimuli are perceived based on their relative difference compared to the background. This is famously known as Weber's Law. From these first three sections, we came to know that all knowledge is based on the five senses and that language is a way of initiating the process of experience.

In the next section, we discussed largely two ways of knowing: science and arts. These are much more complicated and subtle than the mere senses. Humans are the only species in the world to seek knowledge not through their simple senses, but also their intelligent brains. Science is the greatest milestone reached for humankind. Although science builds upon the five senses, it also extrapolates them and produces theoretical models that give rise to incredible amounts of new technology and innovation. The study of science is required to understand to improve and facilitate life on this planet. However, science has its downsides too. This is where arts come in. Where science is more of a concrete set of knowledge, arts are rather abstract and elusive. These two polar knowledge systems give rise to almost all of the knowledge we can comprehend in the world. The combination of these two can transcend the limitations of each of them and provide a more holistic understanding of whatever it is you are studying. It is important to not belittle either one of them as they are equally important regardless of what the modern world values.

Other concepts such as education and politics were designed with an approach to create autonomy and self-actualization. Maslow's pyramid of needs suggests that our greatest drive or motivator is to seek freedom and autonomy. Unfortunately, both our education system and the political system are structured in a way to produce enslavement rather than autonomy. Though my solution is radical and may seem impossible to apply to the practical world, it is only a prototype. There are many political scientists and social workers that have evidence-based models to create a better and more practical social change. However, my writing is not to produce a change in society, rather it is to produce a change in thought. As long as this perspective of social change changes some thoughts about society and education, then I will be content.

The final section focused on the feeling aspect of life. Emotion, Compassion, and memory are all psychological dimensions of life. There is far too little written about them in this document and they deserve a better look at them through other scientific authors. My job was to outline the general principle of feeling in life. I summarized that all emotions are indicators of desire. We feel a particular way because we have been motivated by the hedonic principle. Yet, in reality, those exact emotions give rise to unrealistic expectations and thus we suffer. Emotions are not only psychological, they are also physical in the way they affect humans. Emotion is best defined as a psychological feeling that arises due to a physical sensation. Since we can't change our biology or our physiology, our best bet to control emotions is to control our psychology. When viewing emotions mindfully without any judgment or attachment, they seem neither pleasant nor unpleasant. They seem like waves passing through our minds, yet the individual's mind does not perceive them as good or bad. The individual only observes the feeling and lets it go. In such a stage, the individual finds deep compassion for himself and others. Not only is this spiritual finding beneficial to the individual's journey, but it's also beneficial to society. If all emotions are channeled through compassion and not expectations or desires, everyone in the social benefits. There is some hope of reaching such a utopia if each individual looks at the world with a selfless attitude.

At last, we arrive at free will and the possibility of a god. Regardless of what religion or culture one is part of, they all agree that life is short and there are only certain things we can do. There is no debate about which religion is better than the other. I'd propose that we come up with an assimilated culture where all people come together to purify their minds and serve one another. Religion's basis shouldn't be on hoping to reach heaven or hell, but to create a heaven on earth for all those beings who follow the path of compassion. In reality, we won't have real free will until we can calm our minds and instincts and act purely according to our true nature. Our true nature does not reside in our biology or evolutionary instincts, rather it resides in transcending them and finding a state of mind where no entity disturbs our peace. Our true nature is nothing but peace. Upon reaching it while alive, there won't be any desire to go to heaven in the afterlife, for you have found heaven within yourself in this life itself. Heaven and hell are not places that souls visit, they are mindsets that an individual possesses. A man goes to heaven if his mind is corrupt and full of ignorance, while a man goes to heaven if he has realized the truth and found inner peace within himself. There need not be any search outside for particular heaven or hell. What exists is within us, all of us.

Enlightenment

Five hindrances keep us from attaining peace, and almost everyone would agree with them. They are: 1) seeking sensual desires, 2) Having aversion towards anything, 3) Laziness/ pessimism, 4) Restlessness/ Non-mindful, and 5) Skepticism/ Doubt. Seeking desires or not seeking desires both are seen as a hindrance because in reality when there is a choice taken between wanting and not wanting, you are acting for the sake of the body, which is only a part of you. Having a lazy attitude, or a nonmindful attitude can also be a hindrance. When we are lazy, we let inertia take over and fall into the feedback loop of maintaining that relaxed posture for as long as possible. When we are restless, we also let inertia take over and tall into the feedback loop of maintaining that aroused state of mind as long as possible. If both of these were to be done mindfully, laziness would turn into meditation and restlessness would turn into mindful concentration and both of those hindrances would go away. Finally, a skeptical mind is the greatest hindrance to the practice. Regardless of what spiritual path you follow, or what you truly believe in, if you are constantly skeptical of everything, you will build a habit of rebellion and never understand anything that lies outside your intellectual capacity. By being open-minded and realizing that one's perspective is limited to his or her experience of life, one can openly try new experiences and then have a judgment about them. Being skeptical about any philosophy or spiritual practice before trying it will only lead to the demise of the individual.

What is true enlightenment? After all the knowledge we discussed in the earlier sections of the book, and all the knowledge you may have gained from reading textbooks, online articles, college lectures, and so forth, what strikes you as interesting and worth exploring? What knowledge gives an understanding of self and the world in a way where everything makes sense? Anatomy and physiology? Behavioral neuroscience? Cognitive Science? Philosophy? What area gives us the most comprehensive understanding of ourselves? The answer is none of the above. In reality, that knowledge isn't even considered knowledge, we call it wisdom. Wisdom is the path of enlightenment, not knowledge. Knowing things can get you a job, a family, and many worldly pleasures, but they won't bring true peace. On the other hand, having wisdom will bring clarity and focus in life. Wisdom isn't gained through studying philosophy or reading scripture. Wisdom is gained through introspection. Our innate nature is to gain wisdom and share it with others. But unlike other knowledge, wisdom cannot be shared through mere woods or senses. Nor can it be shared through logic or mathematical equations. Wisdom comes from experience and introspection. So the only way to know about your true self is to introspect in meditation and go deeper with the breath. All religions have some sort of meditation, and now even science has taken interest in it for its therapeutic effects. However, those are only the side effects. After long hours of meditation, one will awaken to a state of self where the physical dimension will disappear and there will be no feeling of pain and pleasure. The language permits no further detail but it asks us to explore for ourselves. By getting rid of the 5 hindrances, one can concentrate on meditation. A prerequisite to gaining wisdom from meditation is that you must have the intellectual faculty to understand the wisdom. You can teach calculus to a 5-year-old but he will never understand it because he doesn't have the prerequisites or the brain to understand a complicated topic such as calculus. Similarly, you can't expect to gain the highest wisdom until you have the proper mindset. It takes a long time to create that mindset, but once you build that mindset of finding balance in each activity and being mindful regardless of what happens, you will receive wisdom from within.

    The bottom line of this chapter is that you must at least try to explore things outside of the physical dimension. Just because our biology can only perceive certain features of the world doesn't mean other features don't exist. By being open to spiritual practices and having a sincere attitude of exploring your true nature, you too can find inner peace and content. But it starts with curiosity and wonder. It starts with the question who am I and what am I doing here? Meditating and reflecting upon that itself will reveal many mysteries about yourself.

I thank you for reading this very informal writing and I hope you gained both knowledge and wisdom along with your reading. I wrote this book for myself, but I feel like my words can reach many people and can help them think about something they never thought about it. The final chapter summarizes the entirety of the book as well as initiates a discussion on the topic of spirituality. It is best to end this book here- nothing seems to be left out.

Preface

This book was written with the intent to explore various areas of knowledge that we use to come to conclusions about ourselves and the world. From the recent existential literature, we realize that few authors have gone out of their way to examine their means of gaining knowledge. I think we must become skeptical about the means before concluding. In my journey as an amateur philosopher, a neuroscience student, and a curious reader, I have realized that my views and perspectives are always changing based on how I intake information. For instance, I can either contemplate the content, I can write/ read it, or can listen to various credible people talking about it. I can look at the piece of knowledge through an artistic view, perhaps a mathematical concept, or a moral perspective. These different perspectives construct our reality and if we don't analyze which ones we use, then we are bound to fall into our own habitual and biased thinking traps. By analyzing how we come to knowledge, I have learned that there isn't one way to gain knowledge, instead, there are many different approaches. We can either isolate ourselves to one area, or we can look at the picture from the top and realize its diversity.

Regardless of what lens we choose to see the picture, we have to understand the limitations of our perception and cognitive tools. Neuroscience and psychology are pioneering fields that are exploring various types of biases and unconscious agendas that hinder us from reality. Understanding the mechanisms of how our minds change our perspective towards ourselves and others will give us a chance to better prepare for when we get in those situations. And the best way to do this is through introspection. The more we introspect on the means of knowledge, the closer we get to knowledge itself. The Socratic seminar is an excellent way to question our beliefs and wonder whether we do know the things we claim to know. Just like we can't use a knife to drink soup, or a spoon to cut vegetables, we can't use the wrong methods to understand the truth of reality. The natural

Preface

question then becomes what is the 'right' method of knowing, and that is still a mystery, but the closest answer to that, at least in my opinion, would be meditation. Meditation/ contemplation on both practical and abstract issues, and trying to find an answer that is unhindered by the senses. This doesn't mean the senses are flawed completely, but their use can be limited to various degrees.

I hope you enjoyed reading this, and that you continue exploring yourself and the world around you. May you find satisfaction in the journey to the truth and help those around you.

Nature of Reality

About the Author

Sattvik Basarkod is currently a junior at Wayne State University studying Neuroscience and Health Psychology, with a concentration in Pre-medical health sciences. At the time he started the book, he had just finished high school at International Academy East and received his IB diploma. His passions include reading, writing, meditating, going on nature walks, and talking to new people.

"All of us have those moments when life seems too boring, and there is nothing to do but indulge ourselves in various activities that distract us from boredom. But I realized that this boredom could only be eliminated if I yearn for something bigger, something that would evoke the strongest feelings within me, something that would touch people's hearts and souls and question their very existence. I, myself have gone through many obstacles that I thought I would never cross. Yet, with strong intention and diligence, I persisted in my endeavors. If anything, I learned there is no point in living if I am not giving my best every day. I wrote this book to reach out to those who have fallen down, who have no support in life, who may be amidst an existential or mid-life crisis, and have no sense of a future purpose/ goal. To all such people, I send wishes of patience and determination. I hope that your journey towards the truth is more valuable than reaching the destination itself. I hope that you can use the suffering in your past to motivate your present self to strive, persist, and reach the goal. And then once you reach it, I hope you share your story to uplift the people around"

Recommended Books/ References

Philosophy
- Republic by Plato
- Metaphysics by Aristotle
- Meditations (Part 1-3) by Descartes
- Meditations by Marcus Aurelius
- The Plague by Albert Camus
- Beyond Good and Evil by Nietzche
- Man's Search For Meaning By Viktor Frankl

Neuroscience
- Principles of Psychology by William James
- The Man Who Mistook His Wife For a Hat by Oliver Sacks
- Siddhartha's Brain by James Kingsland

Buddhism
- Mind Illuminated by John Yates
- Where Buddhism Meets Neuroscience by Dalai Lama
- The Art of Happiness by Dalai Lama

Nature of Reality

Notes

Nature of Reality

www.ingramcontent.com/pod-product-compliance
Lightning Source LLC
Chambersburg PA
CBHW071414210526
45465CB00001B/378